Do Abominable Snowmen Of America Really Exist?

by Roger Patterson

Fourth Edition Reprint
2018

Last Published by
Pyramid Publications
in 1996

ISBN-13: 978-0-88839-079-0 [trade edition softcover]
ISBN-13: 978-0-88839-096-7 [trade edition hardcover]
Copyright © 1966 Roger Patterson
Copyright © 1995 Rene Dahinden

2018 original reprint by Crypto Editions
(an imprint of Hancock House Publishers)

Printed in the United States

Published simultaneously in Canada and the United States by

HANCOCK HOUSE PUBLISHERS LTD.
19313 Zero Avenue, Surrey, B.C. Canada V3Z 9R9
(604) 538-1114 Fax (604) 538-2262

HANCOCK HOUSE PUBLISHERS
#104-4550 Birch Bay-Lynden Rd, Blaine, WA U.S.A. 98230-9436
(800) 938-1114 Fax (800) 983-2262
www.hancockhouse.com sales@hancockhouse.com

DEDICATION

. . . To the young at heart who seldom say "impossible."

. . . To the adventurer who doesn't stop at the foothills, but penetrates deep into the forest.

. . . To the individualist who has enough fortitude to stand up for what he thinks is right.

. . . . To all of those who seek the truth no matter what the cost.

ACKNOWLEDGMENTS

I wish to humbly thank Ivan Sanderson, who has worked long and hard in his search for the truth in this matter. Without his effort, this book would not have been written. My thanks also to Betty Allen, Al Hodson, Jerry Crew, Pat Graves, Prentis Beck, John Green, Lee Trippett, Charley Erion, Paul Manley, Fred Beck, Albert Ostman, John Bringsli, Dr. John L. Hunter, Joanis Gray and Rod Thorton, whose help is greatly appreciated.

Illustrations by the author are based on descriptions given him by those who have actually seen the man-animal about which this book is written. While there have been no actual close-up photos taken, Mr. Patterson feels that his drawings, based on the details described to him, are sufficiently accurate to bear a close resemblance.

TABLE OF CONTENTS

TABLE OF CONTENTS

INTRODUCTION

In the vast and as yet unmarked mountain regions of America exists a living, breathing reminder of prehistoric days. He has many names. In California's mountainous northern area he is known as "Bigfoot." In the rugged Mt. St. Helens region in Washington he's been known to oldtimers as the "Giant Hairy Ape." In the wilderness of British Columbia, Canada, he is "Sasquatch."

Across the world in the Himalayas, he is "Yeti" or the "Abominable Snowman."

In my mind the perfect description for him would be "Aboriginal Giant." Over the past seven years I have personally talked with many people who have seen these primitive giants face to face and in at least one case been actually kidnapped by them!

I, myself, have tracked them through the vast wilderness of Northern California and parts of Washington. I have seen their seven league strides mash down underbrush and leave inch-deep tracks in a truck-packed logging trail . . . tracks that measure over 17 inches from toe to heel and almost eight inches across the ball of the foot!

Yes, there does exist an American Abominable Snowman. He is not a myth . . . not a hoax.

Come with me, then, as we seek out the facts about this Twentieth Century mystery. Come with me as we travel a rugged road into the unknown . . . a road that will take us from the campfires of Indians and mountain men to the pages of a modern-day newspaper . . . a road so charged with excitement and steeped in mystery that you'll never forget one exciting mile of the way.

Roger Patterson

"He who seeketh long enough and hard enough will find the truth, whatever that truth may be."

—*Roger Patterson, Author*

Chapter 1

CALIFORNIA'S BIGFOOT

A PRIMITIVE LAND REVEALS A STRANGE SECRET

The forest was quiet and motionless except for an occasional chatter from a nearby chipmunk. A hawk circled lazily high overhead, peering down on a primitive land that seemed to almost breathe with excitement . . . excitement that seemed to be lurking behind every tree and bush, surrounding the whole area so that even the dust at our feet seemed to spring alive with every step.

A feeling gripped me different from any I had ever experienced as my partner and I trudged our way up an old logging road on that eventful October day. It was something I hadn't felt on my first deer hunt or as I stepped into the ring to meet a tough fighter or even when the whistle blew to start the most important football game of the season. Even as we rounded a bend in the road and looked out over a wondrous land of beauty, I couldn't help but feel a bit jealous of the giant forest around us. I felt if it could talk, it would tell a story so incredible, so absolutely astounding, that man only recently has had the courage to explore it fully.

What we found on that rugged mountainside was by no means a "first" for man. In fact, this strange story is an old one to the first real Americans, the Indians.

As I gazed around me my mind began to wander back over the years and circumstances that lead me into this heart-pounding adventure.

It all started (for me, anyway) in December, 1959, when *True Magazine* first startled America with a story about the unbelievable "Bigfoot" of the vast wilderness of Humbolt County in Northern California. I know for sure there was one American who was shocked and that fellow was me. After reading the article I was like most of my friends and thought it was so fantastic it was hard to believe. But the more I thought about it, the more interested and excited I became. Could such a story actually have happened in the Twentieth Century?

With that question ringing inside my head, I set about to get some sort of answer. Seeking this answer has, beyond a doubt, been more rewarding than any adventure I can possibly conceive. But wait a minute: I'd better backtrack a little and let you read the fascinating *True Magazine* story for yourself.

THE STRANGE STORY OF AMERICA'S ABOMINABLE SNOWMAN

By IVAN T. SANDERSON

Copyright 1959, Fawcett Publications, Inc.

California is a huge state and an immensely varied one. It is nearly 800 miles long and it contains everything from barren deserts to lush tropical jungles. It is full of oddities and enigmas, ranging from almost-active volcanoes to places where commercially-minded proprietors assure travelers that something has gone wrong with good old reliable gravity itself. California is always good for a story, and the rest of the nation is always willing to indulge in a big laugh at its expense. But there is one story that nobody is laughing at any more. And it may turn into the biggest thing to come out of that fabulous state since they found something yellow out at John Sutter's mill.

On August 27, 1958, a tractor driver named Gerald Crew drove out to his job, which at that time was working with the crew pushing a new lumber-access road into the uninhabited and only loosely surveyed territory near the borders of Humboldt and Del Norte Counties, in northwest California. Jerry Crew is a native of Salyer Township in Humboldt County. He is an active member of the Baptist Church, a teetotaler, and I have talked to enough people up there to state flatly that his reputation for honesty, level-headedness and just plain common sense is an excellent one.

The area where this road was being built is, surprisingly enough, an almost trackless wilderness. It is bordered by the Pacific Ocean

11

on the west and Oregon on the north; Highway 299 runs along its southern border, and it stretches some 130 miles inland to highway 99. It is crossed by one winding blacktop road and some lesser roads, plus an assortment of logging trails and "jeep-roads" which are used very rarely. While California is thought of as a heavily populated state, this particular section—encompassing some 17,- 000 square miles—has no known inhabitants at all. Almost anything could be living there, and nobody would be the wiser.

If this sounds far-fetched, remember that it was only a relatively few years ago that a Stone Age Indian named Ishi—the last survivor of a race of men long believed extinct—walked out of the forest and surrendered himself to a startled slaughterhouse butcher in Oroville, California. Oroville is only about 65 miles north of the state capital of Sacramento, and a good deal closer to civilization than the remote spot where Jerry Crew found the first evidence of the "Snowman."

Crew's crawler-tractor had been left overnight at the head of the new road, about 20 miles north of the point where it branches out from the narrow blacktop that runs through the Hoopa Indian Reservation from Willow Creek to a place with the delightful name of Happy Camp up near the Oregon border.

Jerry was an older member of the crew employed by a Mr. Ray Wallace, sub-contractor to the firm, Block and Company, which had contracted with the Public Works Department on behalf of the National Parks Service to build the road. Jerry is a local man. Most of his fellow workers were also local men and included, among others, a level-headed young nephew, James Crew, and two experienced loggers of Hoopa Indian origin.

The country is extremely mountainous. As a matter of fact, in most places it is almost vertical, so that you can only go up on all fours or down on your bottom. It is clothed in a closed-canopy of enormous spruce and other conifers beneath which is a spreading sub-canopy of broad-leafed bushes and small trees, while the ground below this is covered with a mat of mosses and beautiful ferns. No matter how high you climb, the dense foliage and mountainous geography make it impossible to see, at best, more than about four square miles.

The road on which Crew was working crawls laboriously up

12

the face of the western wall of a valley that encloses the stream known as Bluff Creek. All along this mountainous trail there are the stumps of vast trees cut and hauled out, and the great dozers and crawlers clank and roar in the hot summer sunlight as they gnaw their relentless way into this timeless land.

Those employed on this work live, during the week, in camps near the road-head. Some have their families with them and stay there all week; others go home to nearby communities on Friday nights and return on the following Monday morning. Jerry Crew's practice was to return to his family over the weekend, leaving his machine parked at the scene of current operations. He had been on this job for three months before the eventful morning of August 27.

What Crew discovered when he went to start up his "cat" that morning was a series of footprints that formed a continuous track to, around, and then away from the machine. Such tracks would not have aroused his curiosity under normal circumstances, because there were three dozen men at that road-head and the newly scraped roadbed was covered with soft areas of mud alternating with patches of loose shale. What did startle him, however, was that these footprints were of a naked foot of distinctly human shape and proportions but, by actual measurement, a whopping 16 inches long!

His first reaction was to consider them a hoax. Crew had heard of similar tracks found by another road gang working 8 miles north of a place called Korbel on the Mad River earlier in the year, and his nephew, Jim Crew, had also mentioned having come across something similar in this area. Being a practical, matter-of-fact person, he felt considerably annoyed that some "outsider" should try to pull such a silly stunt on him.

He told me that he suspected an outsider because, while his fellow workers liked a harmless joke as much as any man, he knew that this job made them far too tired to go clomping around in the dark making silly footprints in the dirt. Then he got to thinking about this outsider, and wondered just how he could have gotten there without being spotted passing the camps farther down the road, and how he had gotten out again, or where he had gone to. Jerry followed the tracks, and that is where he got his second shock.

Tracing them backwards, he found that the tracks came almost straight down an incline at about a 75 degree angle onto the road ahead of the parked "cat," then proceeded down the road to one side, circled the machine, and finally went on down toward the camp. Before getting there, however, they cut across the road and went straight down an even steeper incline and continued into the forest with a measured stride that varied only when an obstacle had to be stepped over or the bank was so steep a purchase had to be obtained with the heels only.

The stride was enormous and proved, on measurement, to range from 46 to 60 inches and to average almost twice that of his own.

Crew's fellow workers refused, at first, even to go and look at this preposterous nonsense that he said he had found; but eventually some of the men, who had to go by that way anyway, agreed to take a look. Some of them, Jerry tells me, "looked at me real queer," but there were others who reacted differently.

All of them, it developed, had either seen something similar in that general area, or had heard similar tales from friends and acquaintances whom they regarded as reliable. Only the Indians present said nothing. Then they all went back to work.

Nothing further happened for almost a month. Then, once again, these same monstrous footprints appeared overnight around the equipment and also farther down the road toward the valley and around a spring. Still, however, the matter remained a purely local affair for another three weeks, the details known only to the men working on the road and their immediate families.

Then, in the middle of September, a Mrs. Jesse Bemis, wife of one of the most prominent and outspoken skeptics among the road crew, wrote a letter to the leading local newspaper, the *Humboldt Times* of Eureka, which said in part: "A rumor started among the men about the existence of a Wild Man. We regarded it as a joke. It was only yesterday that my husband became convinced that the existence of such a person (?) is a fact. Have you heard of this wild man?"

The editor of the paper, Andrew Genzoli, says that he at first regarded this letter with a thoroughly jaundiced eye. But the longer he saw it lying about his desk the more it intrigued him, and finally he decided to publish it.

He expected a storm of derision. Instead, a trickle of confirmatory correspondence began to come in from the Willow Creek area.

Then, on the second of October, "Bigfoot," as the creature was now being called, appeared again on his apparently rather regular cyclical route, leaving tracks for three nights in succession and then vanishing for about five days. This time Jerry Crew was prepared with a supply of plaster-of-paris, and he made a series of casts of both right and left feet. Two days later he took time off to drive to Eureka on personal business and he carried the casts along with him to show a friend. While he was there somebody told Editor Genzoli about him.

When Genzoli met Jerry Crew and saw his trophies, he realized he had some real, live news on his hands. He ran a front-pager on it with photographs the next day. Then the balloon went up. The wire-services picked it up and almost every paper in the country printed it, while cables of inquiry flooded in from abroad.

The first I heard of the affair was a cable from a zoologist in Paris, who sounded slightly hysterical. I have done a good bit of writing on this general subject, and so I get a lot of esoteric cables about all sorts of things like sea-monsters, abominable snowmen, two-headed calves, re-incarnated Indian girls, and so forth, the majority of which I am inclined to look into because the world is, after all, a large place and we don't know much about a good deal of it. But this one I frankly refused to believe, mostly because I rather naturally assumed that the location as given (California) must have been a complete error or misquote.

It is all very well to have abominable creatures pounding over snow-covered passes in Nepal and Tibet: after all, giant pandas and yaks, an antelope with a nose like Jimmy Durante, and other unlikely things come from thereabouts. And it is even conceivable that there might be little hairy men in the vast forests of Mozambique, in view of the almost equally unlikely but more or less hairless Pygmies of the eastern Congo that are there for all tourists to see. But, a wild man with a 16-inch foot and a 50-inch stride tromping around California is a little too much to ask even Californians to accept.

15

Of course, there have been similar reports of a like creature called a Sasquatch from all over British Columbia for a century but then, Canadians are still *sort* of foreigners, and also there are far too many Indian legends mixed up in their stories.

The amazing thing in this California case was that the world press actually took it seriously enough to carry it as a news item. Not so the rest of humanity. One and all (apart from a few ardent mystics and professional crackpots) rose up in one concerted howl of outrage. Everybody connected with the business, and notably Editor Genzoli, was immediately smothered in a storm of brickbats.

In the meantime, however, a number of other things had happened. Most notable among these was the reappearance of Bigfoot the night before the contractor, Ray Wallace, returned from a business trip. Wallace had heard rumors that either his men were pulling some kind of a stunt up in the hills, or somebody was pulling one on them. And he was apprehensive, he told me, because skilled and reliable workers were not plentiful and, under the best conditions, the remote location was not conductive to the staying-power of anyone. He suspected that someone might be deliberately trying to disrupt his operation, and he was determined to solve the mystery.

Now it so happened that Ray's brother, Wilbur Wallace, was also working on this job and he — in addition to seeing the tracks many times—had witnessed three other startling occurrences. These he described in detail to his brother.

First, it was reported to him by his men that a nearly full 55-gallon drum of diesel fuel, which had been left standing beside the road, was missing and that Bigfoot tracks led down the road from a steep bank to the place where it had stood, then crossed the road, continued on down the hill, and finally went over the lower bank and into the bush. Wilbur Wallace found the tracks exactly as the men had stated, and he also found the oil drum at the bottom of a steep bank about 175 feet from the road. It had rolled down this bank after having been thrown from the top. What is more, it had been lifted from its original resting place and apparently carried to this point, for there were no marks in the soft mud indicating that it had been either rolled or dragged.

Second, a length of 18-inch galvanized steel culvert disappeared from a dump overnight and was found at the bottom of another bank some distance away. Third, a tire for a "carry-all" earthmover, weighing over 250 pounds, had likewise been in part carried and in part rolled a quarter of a mile down the road and hurled into a deep ravine.

Even after hearing these things from his own brother, Ray Wallace remained skeptical. However, on his first morning on the job he stopped for a drink at a spring on the way down the hill and himself stepped right into a mass of Mr. Bigfoot's tracks in the soft mud around the outflow.

Ray Wallace is a hard-boiled and pragmatic man, and he was already experiencing trouble keeping his men on the job. Handpicked as they were, quite a few had simply packed up and left. Wallace was convinced now that somebody was trying to disrupt his work, and this made him furious.

In fact, he got so angry he brought in a man named Ray Kerr who had read of the matter in the press and who had asked for a job on the road in order to be able to spend his spare time trying to track the culprit. Kerr brought with him a friend by the name of Bob Breazele, who had hunted professionally in Mexico and who owned four good dogs and a British-made gun of enormous caliber which considerably impressed the local people. Kerr, an experienced equipment operator, did a full daily job. Breazele did not take a job, but only hunted.

Tracks were soon picked up and followed by them. Then, one night in late October, they saw it. They were driving down the new road after dark and came upon what seemed to be a gigantic humanoid or human-shaped creature, covered all over with 6-inch brown fur, squatting by the road. They say it sprang up in their headlights, crossed the road in two strides, and vanished into the undergrowth.

They went after it with a flashlight but the brush was too thick to see anything. They measured the road and found it to be exactly 20 feet wide from the place where the creature had been squatting to the little ditch where it had landed after those two strides. Spurred by this encounter they redoubled their hunting forays, but their dogs disappeared a few days later when they were following Bigfoot's tracks some distance from the road-head. The dogs were

"Both Were Surprised"

never seen again, though a story was told that their skins and bones were found spattered about some trees.

All this was, of course, taken with hoots of derision by everybody, including those in Willow Creek who had not seen the tracks. There was one notable exception—Andrew Genzoli of the *Humboldt Times.* He, accompanied by his newspaper's senior staff photographer, Neil Hulbert, visited Bluff Creek personally, found some fresh tracks, and photographed them. He also found something else. In following the tracks down the road, they came across a pile of feces of typically human form but, as they put it, "of absolutely monumental proportions."

They contemplated going to fetch a shovel and container and taking this back to Eureka for analysis, but it was a very hot night and there was a five hour drive over a dangerous road ahead of them, and, while they did not say so, it would not be surprising if they were a little shaken by the night's events.

Later, however, Ray Wallace also stumbled upon a similar enormous mass of human-shaped droppings. He shovelled them into a gallon can and found that they occupied exactly the same volume as a single evacuation of a 1,200-lb. horse!

Further foot-tracks and other incidents are continually being reported. Later last spring two fliers—a husband and wife in a private plane—were flying over the Bluff Creek area. It was April and there was still snow on the mountain tops, some of which are bare of trees. It is alleged that they spotted great tracks in the snow and that, on following these up, they sighted the creature that had made them. It was enormous, humanoid, and covered with brown fur. I am still trying to locate this couple, but at the time of writing have not identified them.

Other recent reports are more easily pinned down. Among these are alleged statements by two doctors of having met a Bigfoot on Route 299 early in 1958, and of a lady of great integrity who, with her daughter, saw two—one smaller by far than the other —feeding on a hillside above the Hoopa Valley. This lady, who does not wish her name to be used, also told my partner that when she was a young girl people used to see these creatures from time to time when they went fishing up certain creeks leading into the Hoopa Valley, and that she once saw one swimming across Bluff

Creek when it was in flood. She also states that people did not go above certain points in these valleys because of the presence of the creatures.

More pertinent, however, were a positive flood of further alleged discoveries of similar foot-tracks by all manner of local citizenry over a wide area and extending back for many'years. These all came to light as soon as the local press began to take this matter seriously.

A *Humboldt Times* reporter named Betty Allen did some talking to the Hoopa and Yorok Indians about these matters, and added more to the growing picture.

One middle-aged Hoopa man merely reacted: "Good Lord, have the white men finally got around to that?" More information came from a Mr. Oscar Mack, an older gentleman of the Yurok clan of the Klamaths. "The Bigfoot," he told Mrs. Allen and one of my partners, "were run out of this country by the miners in the 1848-49 gold rush. Before that, there were quite a number of of them."

It is further reported—and this by an engineer now residing near Eureka—that in the 1890's, on the Chetco River in what is now extreme southwestern Oregon, a pair of Bigfeet once infested a mining camp, stole food and wrecked equipment. Three men were later found mutilated and virtually pulverized near the camp, but in open ground where they could not have fallen or been fallen upon by anything other than an animal.

There is a strange rider to this story; namely, that the local Indians said that active volcanoes had broken out in the area to which the Bigfeet fled, in the not too distant past, and that all the game had been killed but that the Bigfeet had survived. Similar stories are numerous; I have transcripts of two dozen, many from old newspapers.

The most recent development came to my attention as I was writing this story. On August 16, 1959, two men, John W. Green of British Columbia and Bob Pitmus of Berkeley, California, found more Bigfoot tracks 23 miles up the new road.

They also found a great number of long, dark hairs, ranging from one to 10 inches in length, stuck on the trunks of fir trees at a height up to 6 feet 4 inches above the ground, plus many piles of droppings which contained small bones, fur, and the residue of various vegetation.

So what are we to make of all this? Obviously, there has been something very similar going on, for a century at least, in this general area. Crazy as it may seem, the whole thing cannot simply be dismissed as a lie, and nobody has yet been able to think up any way in which such a thing could have been perpetrated as a hoax or publicity stunt; or how, or why.

Not even a machine, and certainly nobody on stilts, could have navigated the inclines the "wearer" of these feet negotiated. Besides, it altered its stride, dug its toes 6 inches deep into steep banks going up, and thumped its round heel equally far into them coming down. Something left enormous piles of dung, quite unlike those of any bear thereabouts, and something seems to have toted 50-gallon steel drums full of fluid, iron culverts, and truck-tires for considerable distances.

Something in that area makes a high-pitched, whistling growl. I have had this noise imitated for me by three different people who have heard it in the forest; it is always the same but I cannot describe it.

Forgetting the past and all those who have said they have seen the perpetrators, let us examine the possible explanations for these foot-tracks. They were seen and they existed. Of that, there cannot be the slightest shadow of doubt; the first-hand witnesses are too many, too varied in back-ground, and too honest for anybody short of a complete idiot to any longer suggest questioning their statements. There are many very good reasons for stating that the tracks could not be made by a normal man nor by a machine. Therefore, there are only three alternatives. They must have been made either by an abnormal man, an animal, or a creature somewhere between the two.

For the abnormal man theory there is something to be said, though I would like to stress that a story published in the local press to the effect that a 7-foot youngster ran away from a C.C.C. Camp at Salyer in 1933 is not true. I have interviewed the now-retired head of that camp who states that there never was such a boy there but that a tall lad was lost for four days in the forest but was finally found.

21

There is, however, a factually confirmed story from Idaho, dated 1868, of a 6-foot-8-inch Indian-Negro-White man with a chest measurement of 59 inches, named Star Wilkerson, who was shot for a number of murders by one John Wheeler acting on behalf of the community. His feet measured 17½ inches and he was almost completely wild. The casts of the Bluff Creek Bigfoot measure 16 inches and are completely human except the toes are relatively small, are somewhat rounder, and are arranged more squarely to the long axis of the feet than in the average human being.

Thus, it is possible that there could be some runaway human individual of large but not excessive stature and of considerable bulk and weight—for engineers affirm that the prints he left indicate a weight of at least 750 pounds—residing in the large uninhabited area of northwest California. But there is nothing in the records to indicate who such a person might be or where he came from, and the records have been most carefully searched. (And, there is nothing everybody—the press, the public at large, the few scientists who have deigned to read the accounts, and especially the local inhabitants—would like more than to find out that there is such a runaway human being.)

The suggestion that these tracks are made by an animal need hardly be discussed. The only animal that might be suspected is a bear, of which there are some of the Black (Euarctos) in the surrounding areas (but not, strangely, in the Bluff Creek valley) and just conceivably some grizzled-brown Dishfaced (Arcots) bears, commonly called "Grizzlies." The latter is most unlikely. Further, no animal has a human foot.

The third and last alternative is a Humanoid; an intermediate creature between Man and what we call an animal. The existence of any such creature today, is, of course almost unthinkable. Yet the bones of just such intermediates—and of several grades from the bent-legged Neanderthal sub-men to the stunted-brained Pithecanthropines of China and Java and the pygmy Australopithecines of South Africa with their half-simian, half-human anatomy, and the gigantic 8 to 12 foot, human-like ape called Gigantopithecus of southern China—have now been unearthed. There were "giants in those days." Could some have survived? Above all else, could they have ever lived in North America and survived, in, of all places, California? The astonishing thing is that they could well have done both.

We now know there were men on this continent before the last ice-advance. And during the last million years, during which man evolved, many large and less competent mammals crossed back and forth between the Old and New Worlds in the northern hemisphere—the elk, mammoth, moose, brown or dishfaced bears, and lesser folk like the beaver, marmot, mink and others. Why should not have Sub-men or Ape-men that were resident in what is now northern China, have done so too?

And if they did, why should not they, like the moose and elk, have survived the southward ice advances; even if mammoths, mastodons, the giant beaver, and a large lion did not? After all, some Ape-men made tools and they may have made fire. They would have been better equipped to survive the cold, even if they did not migrate southward during the ice ages. If they survived, moreover, to where would they retreat? Obviously, to those highly inaccessible areas not populated by modern man.

And one such place, strangely enough, is the northwestern tip of California.

Before you are tempted to scoff and put this story down, bear a few things in mind. This area extends over 17,000 square miles, and nobody lives there. Apart from the higher ridges and mountain peaks, the ground area is completely concealed from the air by forest. It has never been properly surveyed or mapped. Yet, for all this, the area is well watered, overgrown with berries, full of small game, and never completely snowed in. Though it nestles in the midst of civilization, and is as fertile and livable as civilization, it is completely uncivilized.

Almost anything could be living in there. From the evidence, something is. Will somebody please do something about it before it is too late?

—Ivan T. Sanderson

Chapter 2

BIGFOOT (OH-MAH) GIANTS STILL ON THE PROWL

In the following few years after I read Sanderson's article I became intensely interested and very puzzled by the whole matter. Here a leading research scientist of very high caliber writes an article in a leading men's magazine, giving names and dates connected with a most astonishing and exciting story. Yet, one year went by—then two, three, four years—nothing on the giants of Humboldt County. Not even another magazine article refuting it as a hoax. Why surely, I reasoned, that if it had been a prank or a hoax of some sort someone would have jumped on it.

In the meantime I had found out some interesting things from some old Indians that live on the Yakima Indian Reservation near my home. Some said that they had heard from their Klamath friends that these creatures had always been in this area, but they themselves wanted nothing to do with them as they were considered evil spirits. Others said huge creatures used to be all over the Northwest and some believe they still roam the mountain areas. Now this was encouraging to me, as I have always respected and believed most of what the older Indians have to say. Nevertheless, none of them told me anything new that I could check on here.

About this time I had business in Los Angeles. Realizing that this would be my opportunity I decided to leave a few weeks early and check out Sanderson's story myself. What I did find was even more startling than I had expected.

Rod Thorton, a friend of mine, went along; although he was very skeptical about the whole thing. But he soon changed his mind. We decided to go Highway 99 through Oregon and cut off on Highway 96 just below the border. We stayed our first night in a lively little logging town by the name of Happy Camp. At supper we met the town constable. He was a believer in the subject and told us we could get more information further on down near Orleans, Whoopa or Willow Creek.

We left early the next morning, driving all day, and pulled into Willow Creek about five in the evening, after the longest, most nerve-wracking 90 miles I have ever driven. Logging trucks came roaring around blind curves at us on a road that had no room for two vehicles, thousand-foot cliffs, no guard rails, the ruggedest country I have ever seen in my life! No wonder it is called the primitive area, (map shows our route, page 19).

Bits of information had been picked up along the way, but it seemed to be leading us to Willow Creek, considered headquarters of Bigfoot country. Here we soon learned that three of the good citizens had much firsthand information.

These three were Betty Allen, writer for the *Humboldt Times* in Eureka; Al Hodson, owner of a prosperous variety store, and Jerry Crew, who was frequently mentioned in Sanderson's article. Betty had collected a priceless scrapbook on Bigfoot from newspaper articles over a period of six years, 1958-64. Later on we will go into these as they will give us an idea of how the press feels on the matter. I might add here that all three of these persons are of the finest character, respected highly by all who know them. As far as I am concerned they have told me nothing but the truth throughout our friendship.

Also, they are people who are not gullible or taken in by every wild story that comes along. They have looked into this physically by searching deep into the surrounding mountains for the bare facts and nothing more. They have seen tracks by the thousands that they state absolutely could only have been made by huge giants weighing at least 750 pounds and walking upright! They have also interviewed eye witnesses that have seen these creatures or heard their eerie howls.

There seems to be a 50-50 split between believers and nonbelievers in the area. The believers are the people who have checked into the matter firsthand, such as the three persons just mentioned. The non-believers prefer to sit in their homes or offices saying, "There's no such thing because I don't believe there could be." Now that's alright, I guess, if you happen to be the type that does-

25

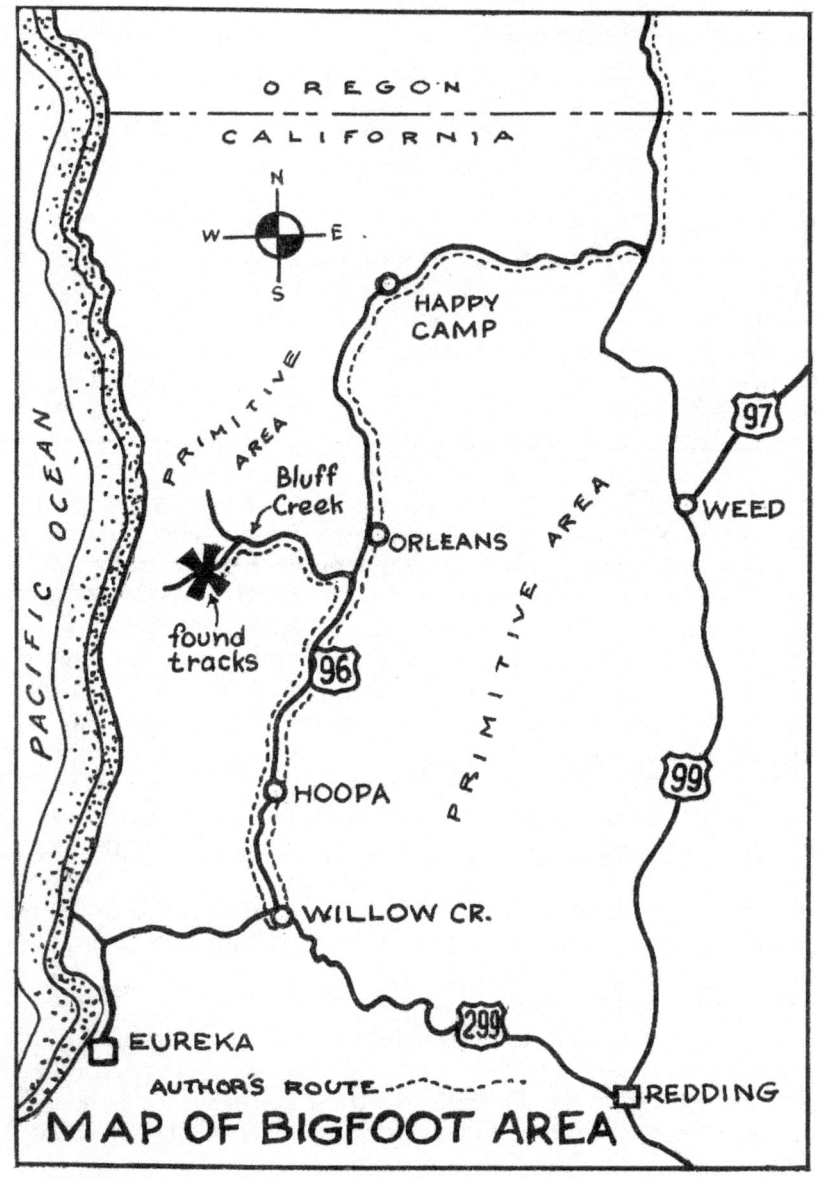

MAP OF BIGFOOT AREA

n't believe something until you see it. But, if this situation is left to that kind of thinking, the matter will never be solved.

As we went here and there gathering our information we learned that Bluff Creek was still the center of activity as there had been some sightings and tracks found. So, of course, that is where we headed.

Just as Sanderson's article has stated, that area is very rugged and heavy with undergrowth, so it makes a good hiding place for anything. Hoping to find fresh tracks, we took along a supply of plaster of Paris

About 35 miles up Bluff Creek we stopped to inspect an old logging landing and to eat lunch. While we were eating, a pick-up truck came down the mountain and pulled to a stop at the landing. A wiry, dark-haired gentleman stepped out and greeted us with a welcome grin. We introduced ourselves to Pat Graves who works for the Forest Service as a timber cruiser. He has been employed by them for the past twelve years and has covered a lot of that country on foot.

Throughout our conversation, we learned that he definitely was a firm believer in the existence of "Bigfoot." He had seen hundreds of tracks year after year, some of them on creek banks thirty or forty miles back from any logging road! Mr. Graves' job takes him to places like this surveying the timber, but to the average hiker or hunter it would be out of the question, as you can't drive any type of vehicle through that tangled brush. The only way is on foot with a pack on your back and at times, he tells us, that's just about impossible too! (Sanderson's article was correct when he stated there were about 17,000 square miles of unmapped area in that region, being only aerially surveyed.)

Pat wasn't always a believer that such things exist. During his first six months on the job, stories from other forest service men and loggers came to him about huge human-looking tracks being found in different places. Before he had seen them he thought they might be those of a deformed bear or possibly someone trying to fix up a hoax. But after seeing the tracks for himself, Pat states flatly, "The thing that makes those tracks weighs around 1,000 pounds, walks upright, takes strides from five to ten feet long (10-foot strides probably running), goes up and down rug-

27

ged terrain where a man just can't go, and has a foot very human-like."

Now Mr. Graves is no fool. He has a good education and is an experienced woodsman who knows every kind of track. He is also a levelheaded fellow who is respected by those who know him.

As we talked more on the subject, Pat said he had seen these giant tracks on Leard Meadow Road by another old logging landing only the day before. We said a hurried goodby and hurried over there. What we found was an amazing sight.

The creature had come down the mountain, crossed a road, gone down around an old logging landing, then over the bank into the brush, taking an average 52-inch stride. The prints were of enormous size—17 inches long and five inches across the heel. I was so astonished I could only stare and try to picture the creature that had made those tracks only the day before. I believe that anyone who sees tracks like Rod and I saw will have to admit there would be no faking them. The imprint of each foot pressed into the ground an inch and a half while our own tracks were barely visible. It was plain to see the foot was flexible as it stepped on small rocks as it traveled down the road. If a rock happened to be where the ball of the foot stepped where the most weight was it was smashed down into the hard road. Where the rocks were up by the toes the foot curled over them like a bare foot would do.

Right here and now is a good time to dispense with the idea that this whole thing is a hoax fixed up by man. In the first place, it would have had to start three or four hundred years ago because some of the old Indians remember their grandfathers telling stories about these giants. In fact, some say these giants were here even before the Indians, and how long ago was that?

Second, these tracks and eye witness reports show up all over the world, Himalayas, China, Africa, South America, North America. In fact, just about on every continent. Some of the sightings are at the same time.

Third, many of the people who have reported seeing these creatures are sound, respectable citizens realizing they will be facing a lot of ridicule by doing so. None the less, they relate their experiences as honest fact.

Fourth, scientists with high reputations, after viewing hundreds of tracks agreed that there would be no faking them because the depth of tracks, length of stride and the ability to climb rough terrain indicates a very great weight that no man or machine could duplicate.

Fifth, in my own case, no one knew that I was coming to that area so why would they bother to put tracks on that logging road?

Sixth, in Pat Graves' case why would someone put fake tracks thirty or forty miles off any road where the chances of anyone finding them would be one in a million?

I'm sure you see that this is no farce, hoax, or idiotic scheme. On the contrary, this could be the biggest scientific breakthrough in the study of the the development of man since the beginning of time. With that said, let's head back to our story.

As we started down the mountain we met another fellow gassing up equipment. All the loggers had gone to their homes in the valley, as it was late in the evening by this time. This man's job called for him to work late after the others had left, so he camped right there in a small trailerhouse. His only company were four or five dogs which he used to hunt bear in his spare time.

The night before he told us, something had come by his place about midnight. He had heard the dogs yelping as if they were scared. He stepped out in time to hear something running off through the woods. Then he went to see the dogs and found them all huddled together, whining like puppies. Now, they were good bear dogs he told us, and this bothered him no end, as he felt sure that the thing had not been a bear. As we left he said he thought he'd quit his job and leave that part of the country. I guess it was just too much for him.

That night we stayed in a small cabin further down in the valley but still in the mountains. Around three in the morning Rod and I were awakened by the strangest sound we had ever heard. It was somewhat of a high-pitched whine trailing off to a deep growl Looking out of the window into the dark night we could see nothing; and we were not about to go out and explore! We could hear brush pop and crack that indicated something heavy was there.

The following morning we looked for tracks. However, the brush was too thick to show any. Had it been a "Bigfoot?" I am not sure. I do know it was the weirdest sound ever to reach my ears.

I promised we would go over some newspaper clippings from that area and some eye-witness accounts from people coming face to face with these giants. Here are some quotes from the *Humboldt Times*, Eureka, California and some very recent articles from the *San Francisco Chronicle*, mainly about the Central Oregon area.

San Francisco Chronicle

THE VOICE OF THE WEST

FINAL HOME EDITION ★

MONDAY, DECEMBER 6, 1965

10 C

The Mountain Giants

A comparison of a man and the "man-animal," compiled from reports based on the numerous eyewitness sightings of the strange mountain creatures.

San Francisco Chronicle, Dec. 6, 1965

Growing Mystery

'Animal-Men' of The Northwest

By GEORGE DRAPER

There is mounting testimony that giant man-animals may be roaming the remote forests and mountains of the northwestern part of the United States.

Two men in responsible positions have told this reporter they encountered the hairy sub-humans at different times and places.

And there exist tape recorded interviews with others who have either seen one of the monsters or heard its hideous scream or smelled what they described as "the foul stench" of its body.

These creatures, they said, appeared to be half ape and half men and weighed more than 500 pounds.

Accounts

A man-animal, according to a Southern Pacific employee named Gary Joanis, picked up a deer he had shot and fled with the corpse into the tall timber.

A Fresno lock and safe company operator said he watched in horror and disbelief as one of these freaks of nature attacked his hunting partner.

Yet another man, who works for the University of Oregon, said he saw one of the giants stride across a meadow.

And a geologist, R. A. E. Morley, has claimed in a lengthy, tape-recorded statement, that someone hurled a boulder at him while he was swimming in a mountain stream and he thinks it was one of the elusive man-animals.

At the Fence

Giants have been sighted as far south as Pinecrest in California, where the so-called "Terror of Tuolumne" was reported last year.

And at Fort Bragg three years ago, Robert Hatfield saw a man-like monster peering over the backyard fence at 4:30 a.m.

All of these sightings, backed in many cases by such secondary evidence as huge, human-like footprints, massive droppings, strange screams and other details, mesh to some extent with world-wide reports of giants.

By far the most frequently mentioned, of course, is the so-called Abominable Snowman of the Himalayas.

Other Giants

Whether this creature exists or whether the enormous tracks found in the Himalayan snow fields were the work of leaping mountain foxes is a question still under investigation.

Other giants reported in recent years include the Daghestan Wild Man of the Caucasus Mountains; the ghost monster of Nyasaland's Zomba rain forest known as the Ufiti; the Tokoloshe, which is believed to have bewitched and seduced a woman in Southern Rhodesia five years ago; and "The Thing," a giant that popped up in the Malayan jungles in 1956.

Although none of these man-animals has been photographed, captured live or shot, there is an eerie persistence about the reports of their existence.

Consequently, a young electronics engineer named Lee Trippett from Eugene, Ore., is making plans to hunt a man-animal of the Pacific northwest by using a radical technique.

In a nutshell, Trippett hopes to influence the lonely creature's subconscious mind by resorting to parakinesia and extra-sensory perception.

"This involves setting up a certain sense of sympathetic vibrations," Trippett said.

There have also been frequent reports of man-animals in British Columbia, where the creature is known as Sasquatch.

John W. Green, publisher of The Advance at Agassiz, B. C., has just compiled

31

a report listing 120 Sasquatch incidents ranging from sightings and attacks to the discovery of tracks and various strange occurrences.

In 1942, Green reports, a man at Katz, B. C., had his arm broken by a hairy giant while picking berries.

The other Green items include:

The Indians at Bishop's Cove, B. C., were reported in 1907 to be terrified by a monkey-like wild man "that digs clams at night and howls."

Henry Charlie was chased more than a mile by a hairy giant near Harrison Mills, B. C., in 1945.

In the early 1800's, two mountain hunters in British Columbia were killed by a creature "that walked on two legs."

A Capture

A train crew captured a half-man, half-beast near Yale, B. C., on July 3, 1884.

"It had forearms longer than a man and could break a stick no man could break in the same way," according to Publisher Green's report.

A list of 41 man-animal incidents reported in the States of Washington, Oregon and California has been compiled by young Trippett.

These include many reported sightings of the beast and the discovery of two sets of tracks in the wilderness of northern Humboldt county.

One set of tracks, each footprint 16 inches long, was found on Bluff Creek on Oct. 4, 1958, by Jerry Crew, field director of the Child Evangelism Fellowship.

Bill Chambers, at that time a reporter for the Humboldt Times, inspected the prints found by Crew and another set of giant prints found in the Bluff Creek area by contractor Ray Wallace.

"Tell you the truth," said Chambers, who now runs a Eureka sports shop, "I don't know what to think of those tracks.

"But it kind of makes a believer out of you when you're way out there in those mountains and follow those tracks at night with a flashlight."

Foot

California agricultural inspector R. P. Doran, who is stationed in the remote and rugged Siskiyou Mountains on the California-Oregon border, said he has heard reports of the Big Foot, as it is known in that area, for many years.

"If it is a hoax," said Inspector Doran,

"then somebody has put it over pretty good."

The Siskiyous at dusk have a wistful, haunting quality. Soft ground fog hangs in the steep ravines, veiling the stands of pine and fir. It is much the same in the Trinity Mountains to the south.

Calvin Rube, a Yurok Indian, stands by his home atop a lonely mountain overlooking the Klamath river.

"The Big Foot was known to my people as the traveler or the patroller. It's his job to keep nature and men in harmony," Rube said.

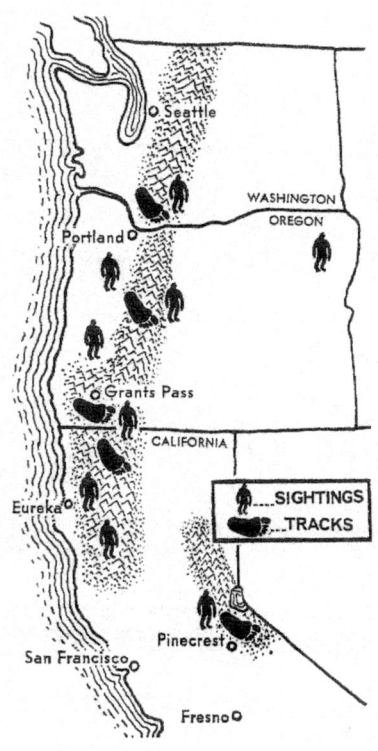

Tracks and sightings of man-animals have been reported from 54,-000 square miles of wilderness in the Northwest, indicated by shaded area. They have also been seen in British Columbia.

32

Although he has never seen Big Foot, he said, he has heard Big Foot's weird cry on two occasions.

"One time," he said, "it made a pale face jump clean over the campfire.

"The other time, we found his track the next morning. It was 18 inches long. And before the track there was a sapling that had been broken off 10 feet above the ground.

Tomorrow: Face-to-face encounters with the man-animal.

San Francisco Chronicle, Dec. 7, 1965

The Mountain Giants

THEY SAW THE MAN-ANIMAL
By GEORGE DRAPER

The belief that several hundred giant man-animals are roaming the wilderness areas of California, Oregon and Washington was expressed yesterday by the owner of a Fresno lock and safe company.

"There is absolutely no doubt in my mind that they exist," said O. R. Edwards, and "and I know they are extremely strong and very intelligent."

Edwards, 56, a legitimate safe cracker with an uncanny knack for twirling the dials of the most formidable strong box, also believes these man-animals are "extremely dangerous" and "capable of mutilating a human being."

"I wouldn't be here today if I'd shot at those I saw," he said with disarming frankness.

Edwards is one of a dozen or more persons who claim to have had face-to-face encounters with the hair-covered giants that they say stand between seven and ten feet tall and weigh more than 500 pounds.

Another terrifying face-to-face encounter with a man-animal has been reported by Don L. Hunter, head of the Audio-Visual Department at the University of Oregon.

Hunter caught a glimpse of a giant taking enormous strides across a meadow at Todd Lake in the rugged Three Sisters Wilderness Area of Central Oregon.

"Even now the back of my neck gets cold when I think about it." Hunter said while sitting in his comfortable home on a hill overlooking Eugene.

33

The university official said he and his former wife, who was with him and who also saw the creature, became so petrified they spent the night in their car after driving ten miles away from the scene.

Hunter said he has been trained in the scientific method of testing everything he sees and hears.

"I can't believe it was an hallucination. Two of us saw it. My wife was just as scared as I was," he said.

Edwards, who said he was "raised in the woods" and "used to be a good mountain man," was hunting with his friend, Bill Cole, when he encountered a man-animal in the southern Siskiyou mountains during World War II.

This is Edwards' account of the meeting:

"We were both moving slowly and quietly around this patch of brush. Bill went around the left side; I was on the right.

"I was sweeping the area ahead with my eyes. On one sweep I caught a glimpse of what seemed like an ape-like head sticking out of the brush.

"By the time I had brought my eye back to focus on the spot it was gone.

"Then I heard the 'pad-pad-pad' of running feet and the 'whump' and grunt as their bodies came together.

"Dashing back to the end of the brush I saw a large man-like creature covered with brown hair. It was about seven feet tall and it was carrying in its arms what seemed like a man..

"I could only see legs and shoes. It was heading straight downhill on the run.

"I was about 30 feet away and the opening in the brush was only ten to 15 feet wide. At the speed he was going it did not leave me much time to make observations.

"I, of course, did not believe what I had just seen. So I closed my eyes and shook my head to sort of clear things up.

"I looked down the hill again in time to see the back and shoulders and head of a man-like thing covered with brown hair. It was disappearing into the brush some 70 to 80 yards below."

To confirm his recollections of what happened that day, Edwards wrote a letter to Cole at Grand Island, Neb., last year.

"I guess I should have started looking for you. I don't know why, but I didn't," Edwards wrote. "Maybe I was afraid of the creature. Maybe I was afraid I'd find you dead."

Cole replied, acknowleging the incident, but saying, "I don't think he packed me at all."

"I was conscious all the time. I didn't hurt any place. Only my conscience was hurt," Cole wrote.

Cole said that after he "quit rolling," he went back up the hill and got his rifle.

"I stood there some time and looked and listened. I had a feeling I was being watched and hunted," Cole added in his letter.

Sitting in the back room of his Fresno lock and safe shop, Edwards said he has heard many stories about the Bigfoot giants from men who claim to have seen them.

"But Cole," he said, "is the only man I know who has had physical contact with a giant."

Edwards mentioned one other thing in his letter to Cole—that for 20 years "I would not believe what I had seen."

And to this, Cole replied:

"Funny, neither of us has guts to say what happened to us."

Prior to the exchange of these letters, it would seem, Edwards and Cole never discussed what took place on that unusual day.

Edwards also said he saw two more man-animals in the brush at the bottom of the ravine as he worked his way back to the logging road where their car was parked.

"Then came the damnedest whistling-scream that I ever heard from right behind me," he said.

"My hackles went up as I whirled just in time to see a flash of something brown disappear behind the tree."

Other observers have described the man-animal's strange cry as "a vibrating sound" or like the sound of a steam locomotive's whistle or the sound of metal tearing.

"All I can say is that it's one of the weirdest sounds I ever heard in my life—a vibrating wail, like a person in pain," said R. A. E. Morley, a geologist.

Morley, who believes it was a giant that hurled a boulder at him while he was swimming in a mountain stream, said he heard the man-animal scream one night in the Siskiyous southwest of Grants Pass.

The sound unnerved and completely deranged a dog, Morley said, and caused the animal to froth at the mouth and hide under the cabin mat.

Edwards said he saw the creature carrying what looked like a man down the hill—'I did not believe what I had seen.'

O. T. EDWARDS
His friends verified it

DON HUNTER
His wife saw it, too

San Francisco Chronicle, Dec. 7, 1965

THE WOODSMAN'S MAN-ANIMAL PICTURE
IS THIS A DREAD MAN-ANIMAL?

A set of fuzzy photographs of what is purported to be the monster man-animal roaming the Pacific Northwest wilderness were uncovered yesterday in a San Francisco Camera Shop.

If authentic, the photographs would be the only ones ever taken of the hulking ape-like creature whose hideous screams have terrorized a score or more outdoorsmen from California's central Sierra to the forests of Washington.

The photographs were brought into the camera shop for processing more than five years ago by a grizzled woodsman who told a "wild story" of being stalked by a hairy monster in the Three Sisters wilderness area of central Oregon.

The woodsman gave his name as Zack Hamilton and he never returned for his finished films.

Dick Russell, assistant manager of Brooks Cameras, 45 Kearny Street, said he was reminded of the old woodsman's eerie tale and the unclaimed film by the Chronicle's account last week of the current search for the giant man-animal.

Russell said he had first examined the film three years ago and "I got prickly all over when I realized they were the pictures the oldtimer said he had taken in the brush. I never saw anything like them."

But he put the photos back in the files, he explained, because "they didn't belong to us."

Then yesterday, he got them out again and delivered them to The Chronicle.

"If Hamilton hasn't claimed them after five years," he added, "I don't think he ever will. I have no idea where he is. I'd never seen him before."

(See circle) If authentic, his photos would be only ever taken of monsters

38

But between his memory of the woodsman's vivid tale, and the photographs, however ill-focused, Russell said he was convinced they were the real thing.

An old woodsman couldn't fake a thing like that," he said, earnestly.

San Francisco Chronicle, Dec. 8, 1965

By George Draper

A 152-pound electronics engineer yesterday outlined his plan to enter the wilderness unarmed and brain-wash a 500-pound man-animal.

He is Lee Trippett, 35, of Eugene, Ore., who believes the murky wastes of a giant's sub-human cranium may hold a clue to man's total nature.

Instead of a rifle, Trippett will rely on the powers of para-kinesia (PK) or mental telepathy and extra-sensory perception.

For his theater of operations, Trippett has selected the rugged and remote Three Sisters Wilderness Area in Central Oregon.

Numerous evidence of man-animal life have been reported from this area and several persons claim they saw the giants in the shadow of the snow-capped Three Sisters.

Trippett, a quiet, unassuming scientist with close-cropped hair and a shy smile, is convinced man-animals are roaming through 54,000 square miles of uninhabited mountains in the Northwest.

"There is a part of man's mental make-up that is missing," Trippett said, "and it can be found predominant in the giants."

"To discover this quality of the giants' subconscious mind could be a tremendous aid to man's understanding of his total nature."

Trippett and his father, Ben, the operator of a building maintenance business, believe the giants have developed their sub-conscious minds to an extraordinary degree and use this for protection.

"His (a giant's) PK is so great," Ben Trippett told this reporter, "that he can terrorize you from the far side of a mountain."

The Trippetts also believe the giants have highly-developed extra-sensory capabilities which would enable them to receive relatively weak mental telepathetic messages from man.

"My plan," said Lee Trippett, "is to establish a rapport with this beast on the basis of mutual respect and love."

While living in the wilderness "like a child of nature and without artifacts," Trippett plans to stage deliberate meditation or broadcasting sessions to reach the man-animal.

Trippett said the best time to meditate on a giant's ESP wavelength would be during the full moon.

"This is the period when the creature has the greatest confidence whereas man's is when the sun is shining," he said.

If the giant receives the love signal. Trippett went on, and knows that the man broadcasting it is devoid of fear or hostility then there will be a chance for communication.

This reporter has had access to a voluminous file of correspondence between the Trippetts and others concerning contact with the man-animals.

One undated letter written to a person with proven ESP capabilities states:

"Do you recall if you sensed an evil nature about this creature or was it more the absence of the spirit? Also, what chances do we have of mentally meeting this thing without serious repercussion? (It reportedly has very offensive and powerful PK abilities) To what extent must we develop this 'love' approach you suggest?"

Ben Trippett, who maintains a kind of command post in the office of his ranch-type home where he coordinates the search for the man-animal, claims the quarry they are seeking is "slicker and cagier than the bobcat or the cougar."

He also runs something called the Western Research Foundatain, P.O. Box 1202, Eugene, Ore., which is trying to photograph one of the giants and obtain tape recordings of his screams.

The older Trippett, however, is somewhat cautious about the idea of a giant responding to a loving message on his ESP and visiting their lonely camp.

"This question is: What will you do when you see him?" he asks. "He will be so over-powering you may be frightened out of your mind."

Both Trippetts agree giants are so elusive and have so successfully evaded photographers and hunters because of their highly developed ESP.

But the fact so few giants have been sighted does not discourage them.

The Giant Panda was believed extinct at one time, they said, because none was seen between 1869 and 1915 in Western Szechwan, China.

And, right in their own State of Oregon, they said, a 28-pound wolverine was shot on September 11 of this year.

It was the first recorded killing of a wolverine in Oregon since 1912. The mean little beast was thought to be extinct, they said.

The following is a tape recorded interview recorded by Lee Trippett with Don Hunter. The interview was recorded October 18, 1963. Mr. Hunter has been head of the Audio Visual Department of the University of Oregon since 1946.

Hunter—My wife, who passed away in 1951, and I used to go up into the Three Sister-Cascade area every year for vacation. This year, 1942, we went up quite late in the summer as was often the case. The war was on and there were not many people around because of travel restrictions and permits needed to get in the area. They were afraid of fire and sabotage. In this particular area we did not need permission. We had been, I think, at Big Lake and spent the night there.

The next day we went up to the mountain area exploring, taking pictures, and the like. We came to Todd Lake, which is a little ways north of the Century Drive, just before you come to the Sparks Lake. There was a forest camp there, but at that time it was not very well used and it was quite primitive. We arrived there around 3:30 or 4:00 in the afternoon.

As we got there I think we got out of the car and started looking around for a place to make our camp, when a rain storm came up. We sat in the car to weather out this storm as it was raining pretty hard. I looked out across Todd Lake. There is a meadow across the lake and, right in the middle of the meadow area

41

there was a tall figure. It was just standing there. I pointed it out to Delores, my wife, and said, "My goodness, what's that?" It was difficult to see clearly from inside the car so we got out to get a better look. In getting out of the car we must have been heard because it took off for the trees with giant strides. (The situation was illustrated with pencil and paper.) I thought it could have been an elk or something seen head on, but there it was striding with 2 legs. There were no 4 legs about it. When he disappeared into the woods we became petrified. We thought it might come toward us, so we got back into the car and took off. We drove 8 to 10 miles away and spent the night in the car. It continued raining most of the night.

Trippett—How far away do you suppose it was?

Hunter—Well, it was across Todd Lake and in this meadow. I would say, oh, the lake is not more than a quarter of a mile.

Trippett—Was it big enough that you could recognize the fact that it was humanoid in form?

Hunter—Yes, right. It was standing upright.

Trippett—Did it seem as though it were skinny or slender?

Hunter—Yes, it seemed to be skinny. It was not humped over but very erect, with a military bearing.

Trippett—Did it seem as though the arms were very long or out of proportion or that it might have had a very short neck?

Hunter—No, I do not remember that, except that the legs seemed very long. He did not run, he walked. Giant strides very quickly.

Trippett—Were you able to tell whether the thing was covered wth hair or had any clothes?

Hunter —No, I could not tell from that distance. I really do not remember.

Trippett—Did it give off any noise?

Hunter—No, there was not a sound.

Trippett—Have you in your travels in that country seen any unusual tracks or droppings?

Hunter—No. When I had reported this to the Sisters Ranger Station I asked if there were any other reports of a tall man in the area. I remember coming away from there rather discouraged. They seemed to think I might be nuts.

(In here there is a relating of many of the reports for the local area.)

Hunter—Yes, this hearing, too, of the events you have been relating to me has certainly helped explain the unexplainable. It seems that every time I think of this or talk about it to others, which hasn't been often, it would make the back of my neck cold. Let me know if there is any way I can be of assistance. I would be interested in following this up and if there are any expeditions I know what to look for.

EDITOR'S MAIL BOX
San Francisco Chronicle, Sunday, Oct. 26, 1958

Further investigation into the mystery of Bigfoot is due soon, when the Wyoming Expedition, Inc., arrives here to begin a check of the Bluff Creek country.

Arriving here this weekend in advance of the expedition were Hank Alberts of Douglas, Wyo., and Rich Rogers of Weiser, Ida., who have been checking stories about Bigfoot, and visiting the Klamath-Trinity country.

Alberts said Sunday that the members of the Wyoming Expedition are all experienced hunters and trappers, and are well acquainted with North American continental outdoor life. Alberts said that if Bigfoot is a reality, it is not their intention to attempt to trap or kill the creature, but rather to observe and photograph.

"Whatever Bigfoot is," Alberts said, "he deserves to live in peace and quiet. He evidently has bothered no one and is apparently harmless and deserves to be treated fairly." Albert added, "However, curiosity and knowledge must be served and we won't hurt him."

Another expedition, Western Expeditions, Inc., headed by Gary Pesek from Aberdeen, Washington, is reported planning to enter the Bigfoot investigation. It was reported that advance men from the expedition had been checking on the story earlier in the month.

Actively engaged in the Bigfoot mystery is the Pacific Northwest Expedition which is headed by Tom Slick, Texas industrialist and philanthropist, and a group of associates. This group has been active for the past month in the Bluff Creek area.

The Humboldt Times

Eureka, Calif., Tuesday, October 14, 1958

HUGE FOOTPRINTS HOLD MYSTERY OF FRIENDLY BLUFF CREEK GIANT

Giant Footprints Puzzle Residents Along Trinity River

By ANDREW GENZOLI

There is a mystery in the mountains of northeastern Humboldt county, waiting for a solution . . .

Who is making the huge 16-inch tracks in the vicinity of Bluff Creek? Are they the tracks of a human or a hoax? Or, are they actual marks of a huge but harmless wild-man, traveling through the wilderness? Can this be some legendary sized animal?

Everyone is asking the question in the Trinity Klamath county from Willow Creek to Bluff Creek itself. Evidently, there is nothing to be afraid of, for the track-maker, be he animal or human, has never been known to molest or frighten anyone. In fact it evidences a curious interest in the activity of workers wherever he travels. The question is: What is this—is there an answer?

The latest appearance of the huge thing — individual or animal — occurred again sometime Wednesday night and early Thursday morning. The tracks long and wide were found in almost the same location they were seen a week ago. The area is in new construction on the Bluff Creek Timber access road being built by the Bureau of Public Roads. The country is some of Humboldt county's deepest wilderness where not all of its natural secrets are known to the white man.

Jerry Crew, a cat-skinner for the Granite Logging Company, working on excavation, brushing and pioneering for the new road, came to Eureka yesterday, bringing with him a plaster-of-paris cast he had made of the big track. The cast had been poured into a soft spot with a two or three inch depression, but the major features of the foot were well defined.

44

Named "Big Foot"

The big foot, as cast, was 16 inches in length, with the wide seven inch width. Crew said the men refer to the creature as "Big Foot."

Crew said the owner of the big feet had come down the steep mountainside, through an old burn. The surface of the ground was shale, until it had hit the soft earth turned over by the construction equipment. The big walker had burned his tracks along the road, moving along for at least three-quarters of a mile before he changed his course and headed off the road and headed in to more shale. This time its path faded away.

Crew said yesterday, that Robert Titmus, a taxidermist from Redding, and an associate from Portland, Oregon, have studied the newest tracks. They were convinced, and they definitely state, that the tracks were not made by an animal. Crew and Titmus are both well acquainted with outdoor life.

Another observer, Raymond Wallace, a member of the construction crew, also made a study of the tracks on several occasions. He found that the normal stride of the big feet was 50 inches. A measure was made of its running stride, in the path of an evidently fleeing deer, measured 10 feet.

What is it? Well, Crew, who has the backing of his neighbors and friends for truth and reliability, says he doesn't know. However, he is convinced the tracks are made by a human being, or something resembling a human being. If it were a bear, he says, in most cases there would be claw marks, as well as other indications. None of these exist.

The maker of the big marks is not a new visitor in the wilderness. About two years ago there were reports of filled 50-gallon gasoline drums being lugged around like playthings. That was while Crew was working with Granite Logging and Wallace Brothers on a job on a timber access road about twenty miles up Bluff Creek. This summer he has seen hundreds of these tracks. So have other men.

The men are often convinced that they are being watched. However, they believe it is not an "unfriendly watching." They leave their equipment out at night and in the morning the tracks are found going right on by without any apparent curiosity. To his knowledge, Crew says nothing in the line of equipment has been touched. Nearly every new piece of work, though, finds tracks on it the next morning, as though the thing had a "supervisory interest" in the project.

Crew says he isn't superstitious, nor the finicky kind, but he is interested enough to want to know just what this is all about. He is planning some kind of photography setup, in hope of taking a photo of the nocturnal visitor.

While the tracks of old Big Foot have been in evident for some time, and "rumors" have been received at the Humboldt Times office, the first new interest came to light with a note from Mrs. Jess Bemis of Salyer, on September 19.

Mrs. Bemis reported that her husband, Jess Bemis, while working on the Bluff Creek job had seen the tracks along with 15 other men. She said: "On their way to the job, tracks were seen going down the road. The tracks measured 14 to 16 inches in length. The toes were very short, but were 5 to each foot. The ground was soft and the prints were deep, suggesting a great weight. The tracks were wide as well as long. Things, such as fruit, have been missed by those camping on the job."

There came reports from other areas of the big footed wanderer. Larry Knudsen of Fieldbrook knew of men working the Simpson timber country about eight miles north of Korbel on the North Fork of the Mad River who had seen tracks in construction work there. Like everyone else, the men, Ken Craig, of Arcata, Jim Echlund, of Korbel, and a number of others were baffled.

Determined to trace the facts concerning the big feet, I called Julian Pawlud of 2822 D Street, Eureka, who was among the men on the Korbel scene to witness the big tracks. The tracks he had seen, he said, were observed last spring, while he, with others, were doing a logging road construction job. He described the tracks as being "pretty heavy" and made by "bird or human." He didn't wish to be carried away by what he had seen, and he referred to the mark as being of a "paw" rather than a foot.

Traveled Over Culvert

He said the Korbel woods impressions were found on the edge of a creek, and following the path, they went over a sixty foot culvert into a freshly graded road. The tracks were sunken enough to indicate the weight of the object. Pawlus said that there were three toes straight out and a couple of smaller on the side.

At the time, he said at least 25 persons stopped to view the unusual tracks.

In a note to the Humboldt Times during the week, Mrs. Paul Keesey of Pepperwood, reminded us of a hoax played in the woods country. She wrote: "I have read with interest the articles on the huge prints. I recall about eight years ago, when up around Trinidad on one of the logging roads, they had a similar thing happening. This was told to me in my restaurant at Bella Vista hill by several truck drivers. When they investigated, it turned out to be the work of a prankster."

However, most of those with which the Times has been in contact, while not desiring to become too serious about Big Foot, are curious, and seeking an answer. Few people feel that there is anything to fear.

"When I asked some Indian friends about the big tracks," a friend wrote, "they merely smiled, and didn't care to comment." Apparently they aren't worried.

Who and what is Big Foot . . . the maker of those 16-inch tracks? Surely there must be an answer. Maybe someone will succeed in finding it.

'BIGFOOT' SIGHTED BY TWO WORKMEN NEAR BLUFF CREEK

The Humboldt Times, Oct. 14, 1958

Two husky construction workers with good eyesight insisted emphatically Tuesday that they had seen Bigfoot whose massive tracks have been spotted frequently of late in the Bluff Creek area of northeastern Humboldt county.

"He—or it—bounded across the road in front of our car Sunday evenin'," said Ray Kerr, 43, of McKinleyville. "It ran upright like a man, swingin' long, hairy arms.

"It happened so fast, it's kinda hard to give a really close description. But it was covered with hair. It had no clothes. It looked 8 to 10 feet tall to me."

His report to a reporter and photographer from the Eureka newspapers was echoed by Leslie Breazeale, 35, also of McKinleyville, who was riding with Kerr.

Bigfoot's prints, it has been claimed, actually have appeared in Del Norte, not Humboldt county, thus placing the puzzling case in the jurisdiction of Del Norte Sheriff Oswald Hovgaard — if there is any need of legal action.

Kerr doesn't think it was human, but also doesn't believe it was a bear or any other animal he has ever seen. "I was raised in the brush, but I've never seen anything like it."

Breazeale, who awoke from light slumber when Kerr slammed on the brakes, saw the thing as it leaped into the brush.

"I don't know what it was, but it wasn't no man—that's definite," Breazeale related over coffee in the remote wilderness camp.

How are they so sure it was the same animal that made the previous tracks in the vicinity?

Tracks in the dusty road were identical with those seen in the construction area. Both men inspected them with the aid of a flashlight. Their sighting took place about a mile and a half below the construction camp, which is only half a mile from the older tracks.

Kerr said an Indian, Charlie Beach of Trinidad, told him he saw such tracks while trapping in the area 20 years ago, but that nobody would believe his story.

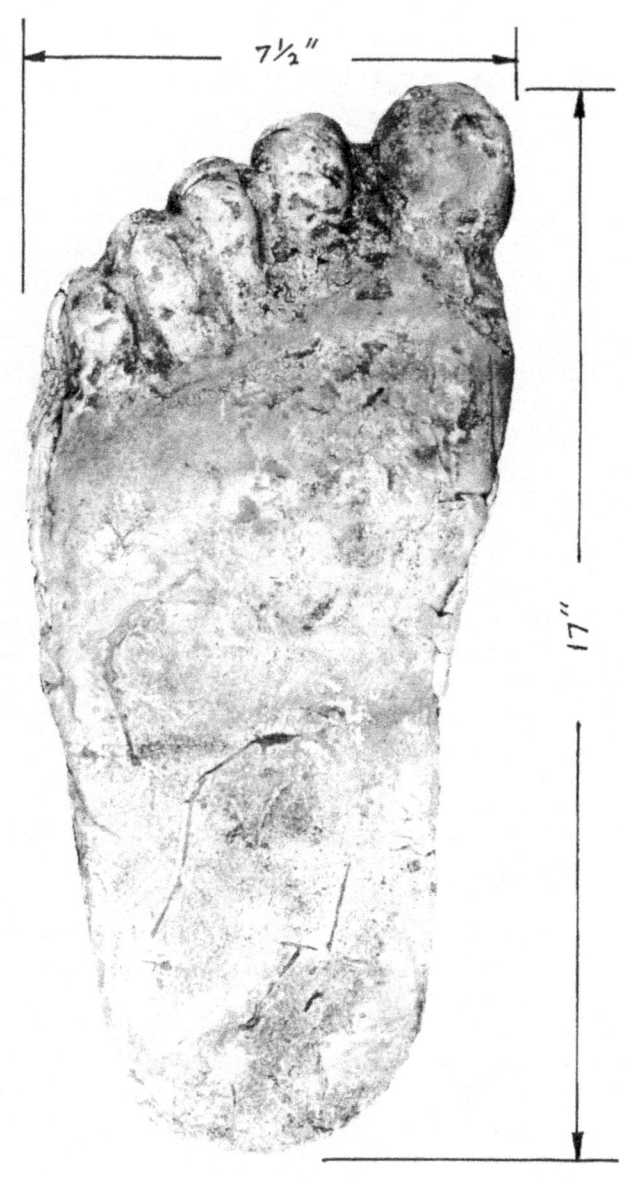

7½″

17″

Cast of "Bigfoot Track," 17 inches long, 7 ½ inches across the ball.
Made October 21, 1964 by Author near Bluff Creek.

An earlier sighting was reported to Roy Wallace, 40, Willow Creek, one of the three brothers who are partners in the firm constructing the road.

This, he said in a lengthy interview, was reported to him by an employee who shortly thereafter left the job and has not returned since.

This description tallied with that of Kerr and Breazeale—tall, hairy, walking stooped over, with long dangling arms and "four feet across the shoulders."

This sighting took place in the early morning hours when the bulldozer operator had just reached his tractor. Bigfoot apparently was drinking from Bluff Creek when spotted, then bounded up a steep incline into the brush.

The chat with Wallace also brought out in vehement terms that he is more than slightly perturbed over accusations allegedly made by the Humboldt sheriff's office that he has perpetrated a hoax on his own construction job.

"Who knows anybody foolish enough to ruin his own business, man?" Wallace asked.

He referred to the fact that about 15 men reportedly have quit their jobs on the project since sightings of the giant footprints.

"I've got three tractors sitting up there without operators, man, and the brush-cutting crew has all quit.

"It just doesn't make sense."

NEW BLUFF CREEK MYSTERY PUZZLES INDIAN: 4 DOGS FOUND RIPPED TO PIECES

The Humboldt Times, Oct. 19, 1958

An Indian who works near the Humboldt-Del Norte county line believes he may have discovered signs of a "Bigfoot" temper fit, a Eureka man told The Humboldt Times yesterday.

Harold C. Goodwin, 66, said Curtis Mitchell, an Indian who works for him, discovered the mutilated bodies of four dogs last Sunday evening.

"He told me they looked like they'd been ripped apart," Good-

They found out "he wasn't a bear!"

win said at his home just off the Elk River road about five miles south of Eureka. "And the bodies were still warm."

The Indian told Goodwin that all of the dogs had been torn apart and one of them had apparently been slammed against a tree. No footprints were found and Goodwin said the Indian "didn't stick around" to investigate after finding the dogs' bodies.

Goodwin, a superintendent for the Sharp Construction Company, has been working in the Bluff Creek area on a concrete bridge about two miles south of the county line for the past four weeks.

A Humboldt county resident since 1898, Goodwin said he "used to think the big footprints were just a joke" but now he's convinced there is some sort of human creature wandering through the northern California wilderness. "I think it's the straight goods."

He added an eerie note to the speculation about Bigfoot: " . . . and the fellow who owned those dogs might be lying up there someplace, too."

The construction worker said the discovery of the dogs has changed the minds of many skeptics working in the area. "This," he said, "is when all of us old-timers start to believe."

Goodwin said he had sometimes been staying on the job over the weekends because it was such a long drive back to Eureka. "I think I'll be coming home from now on, though," he added firmly.

THE FOLLOWING STORY WAS TOLD TO ME BY LEE TRIPPETT

By GARY JOANIS
Recreation Major, University of Oregon, Eugene, Ore.

In the fall of 1957 near Wanoga Butter, Mr. Joanis while with his hunting partner, Jim Newall, saw a giant human-like form come out from some brush not more than 100 feet away. The creature walked over to a small deer that had just been shot by Gary and picked the animal up with one arm and walked very quickly with tremendous strides over the ridge. The creature was described as not less than 9 feet tall with very long hair on the arms. The creature was making a very strange whistling scream.

Mr. Joanis describes more details and knows of one or two others who have seen strange things in this same area.

WYOMING EXPEDITION TO TRY AND TRACK DOWN BIGFOOT

The Humboldt Times, Oct. 22, 1958

Editor's Mail Box
RR5 Langley, British Columbia

Oct. 22, 1958

Editor
The Humboldt Times
Eureka, California
Dear Sir:

I am interested in an article in the *Agassiz Advance* about tracks seen in California. Have you any pictures of these footprints or some descriptions of them that you could send me? I claim to be the only living man that seen a Sasquatch. In fact I was kidnaped by one and lived with them for about 6 days before I got away from them. I claim they are human beings that for one reason or other got left behind from our civilization, but it's one thing about their feet that is not like a human foot today. I saw their feet close up. In the family that kidnapped me was one old man, one old lady, one boy, and one girl. So I know what kind of feet they have, old or young. Your guess about size is about right. It's hard to estimate when you have nothing to compare them with. I always estimated that the old man like I call him would be about 8 feet tall and weight about 800 pounds, a hundred more or less.

I hope to hear from you.
    Sincerely,
        Albert Osterman.

'BIGFOOT' WEIGHS MORE THAN 800 LBS., DECIDES GEOLOGIST

The Humboldt Times, Nov. 18, 1958

Another "Bigfoot" discovery broke in headlines in San Jose yesterday from Los Gatos, where Dr. R. Maurice Tripp, geologist and geophysicist, completed the cast of a 17-inch footprint he found in the Bluff Creek area of Humboldt's storied wilderness.

Dr. Tripp reported that his engineering studies of the soil properties and depth of the footprint from which he made the cast, indicate the weight on that huge puppy to have been more than 800 pounds.

"The print is distinctly different from that of a bear or any other animal known to be in the area," Dr. Tripp told the San Jose News.

The story admitted that although Dr. Tripp did not discount the chance of fraud, he believes that the tracks lend credibility.

"Several people have tried to track him and in one instance his footsteps could be followed a distance of a mile and a half through bush country," he said. "It would be difficult to fraudulently prepare hundreds of such tracks overnight—particularly in the type of country in which they were found."

He mentioned other evidence of "Bigfoot." In the area was a strand of hair found on a tree seven feet, three inches above the ground. Dr. Tripp said the hair was discovered immediately after reports had been received that the mysterious monster was in the area.

Dr. Tripp said his interest is both scientific and the result of curiosity. He first became interested 18 months ago when he received a report that a clergyman and a woman of Bluff Creek said they actually had seen the elusive "Bigfoot." He was able to reach the region in time to get the cast of the footprint.

"Now we just have to find the foot that fits it," he observed.

Numerous expeditions have ventured into the Bigfoot country, at least two of them conducted on a scientific basis. Many others were indivdual efforts, but none has come up with any specific proof either in photographs or by contact with the mammoth mystery man of Humboldt. Indians in the area say the legend goes back to about 1850. Dr. Tripp is the first investigator of scientific note to estimate "Bigfoot's" avoirdupois.

BIGFOOT A LOST HUMAN?

"Has some human being ever been lost in the primitive area near Hoopa?" asks Maggie Hostler of Hoopa. Her question is prompted by a tale she reports told her by a completely competent Eureka resident. The story follows:

"My husband loves to fish and goes often to the remote area down river from Hoopa. He usually carries a camera, but on this day he did not. He had finished fishing for the day and was resting quietly on the hillside. He saw something move. He looked and he saw, standing in the water, what appeared to be a giant man. It was holding two large salmon cross-wise, and in his hands the fish looked like trout. His body, as much of it as was showing, was covered with hair, not like the fur of an animal, but like the hair on a man's head. The hair from his head reached to his waist line. He was about ten feet tall. He stood for a long time in that position holding the fish, just standing there. The man watching never moved, he said he couldn't have if his life had depended on it. When the large man left, or as he left the river, he was gone from sight as there was a bend in the river and he went around it. Bigfoot? I wonder. The further down the river a person goes the more primitive it becomes."

'BIGFOOT' MAKING BIG TRACKS IN U. S. NEWS

The Humboldt Times, Oct. 7, 1958

WILLOW CREEK—With the attention of all the world suddenly focused on Bigfoot and his tracky exploits, the folks along the Trinity River from Willow Creek to Bluff Creek are enjoying the show as much as anyone.

If Bigfoot is within hearing distance, he is either dismayed or pleased, but as of last night he had not made a new appearance. Some speculation has taken place as to whether this individual, creature, or what have you—could be someone who is trying to get away from it all, only to be caught up with.

Bigfoot has been a familiar character to this part of the world off and on over a period of years. While most tales told hereabouts have a legendary flavor, Bigfoot left that category when he began to permit his presence to be known through a collection of footprints.

An ordinary big foot would have been accepted as just that—a big foot. But when a foot soars to 16 inches in length, with a 7 inch width—well, that is something else.

Local experts—and there are a lot of them—are all speculating as to who might be responsible. Everyone admits there must be a "who" attached to this, whether actual prints of a human foot, or possibly those of a manufactured sort.

Mrs. Betty Allen, Humboldt Times reporter for the Trinity-Klamath country, who is known for her factual newswriting, visited the scene and offered some thoughts. She said it was a long trip, but worth making.

"If those tracks are the work of a prankster, he's an artist," she said. "I looked down on at least twenty of the tracks that had been made last Thursday, and they were just as perfect as those made by anyone else walking around the road."

"You could see the toes and the rest of the foot very plainly. There was no exaggeration to be found, that I could see."

Mrs. Allen said that if it is the work of an oversized human being, he is minding his own business and not hurting anyone. She said she had heard some talk about tracking Bigfoot with hounds. "If they do that," she said, "then they are making a mistake. The country he is in is public land and he has as much right to be there as anyone else, as long as he behaves himself."

Law enforcement officers aren't concerning themselves over Bigfoot, because if he exists, he's minding his own business. So far, he hasn't broken any laws to worry anyone.

CANADIAN SATISFIED BIG TRACKS GENUINE

The Humboldt Times, Eureka, Calif.

"I am completely satisfied the tracks at Bluff Creek are genuine," John Green, publisher of the Agassiz-Harrison Advance, Agassiz, British Columbia told the Humboldt Times last night in a telephone conversation.

Green and his wife traveled to Bluff Creek last weekend to see for themselves, the work of Bigfoot, which has baffled scores of Humboldters. Greeen saw the tracks, the size of the strides, the locale they were made in. While there he also interviewed Jerry Crew, Jess Bemis and others who are working on the Bluff Creek project. Green said their attitudes satisfied him that they are telling the truth.

The British Columbia publisher, who has been studying tracks left by the "Sasquatch," a giant reportedly seen in the Agassiz area, came here to compare tracks and information.

Green said on the way home he called on Robert Titmus, Anderson, California taxidermist who had made a close study of the tracks. Titmus discounted the claim that the tracks were of bear origin. Titmus pointed out that unlike the human foot, the largest toes on a bear are in the center of the foot, and clawed. Titmus also said that even the 1800 pound Alaskan brown bear does not have feet the size of Bigfoot.

"It is simply impossible to make so many tracks fictitiously," Green said. "The weight of the one making the tracks had to be extremely heavy to make such deep imprints, even in the hardest of earth." Green said that in nearly every print it was possible to observe the mechanical function of the toes, as they dug into the earth. It is not possible for false feet to make such impressions, for each step varies with the softness or hardness of the earth to which the various toes come in contact.

"Bigfoot is walking a regular beat," Green said, "and it seems likely that he will return. There is an old gold rush trail over which he has been walking for years, and now the new access road cuts right through it, so that is probably why he wanders a bit on new construction. He gets back on the trail and keeps on going. He goes in one direction."

Titmus told Green that he gets 2 to 3 reports a year about Bigfoot. The tracks are not new to the area.

"I know they aren't fake," Green said. "If any one will get out from under his desk and investigate, I think they will agree with me. The trip to Humboldt county was worth making," he added.

BIG TRACKS HAVE APPEARED ALL DOWN THROUGH HISTORY

October 15, 1958

The story of ancient and mysterious giant-sized tracks are an old, old story, Dr. T. D. McCown, professor of physical anthropology at the University of California, said yesterday.

Speaking of Bigfoot's gigantic 16-inch tracks which have appeared on the Bluff Creek access road construction job, Dr. McCown said that such tracks have been reported since the beginning of time.

He said that records show reports of footprints, most of them the same size as the ones found in Humboldt county. Such records indicate that millions of huge tracks have been found through the years. He did not specify their origin.

Most of the track reports have come from Africa and Asia, although many have been reported in North and South America. Some tracks have also been found in Europe, the professor said.

If the Humboldt county Bigfoot is tracked down and discovered, it will be the first time in history that the mystery has been solved, Dr. McCown said.

He went on to say that there have been also many reports of tiny footprints supposedly made by little people.

BOYS SPOT 14-FOOT 'THING': DID BIGFOOT MOVE NORTH?

August 2, 1959

ROSEBURG—It couldn't have been an abominable snowman, because it was raining at the time, but two boys told police here Wednesday that they saw a 14-foot manlike creature stalking through the woods near Ten-mile, about 15 miles southwest of here.

In fact, one of the boys took five shots at "the thing," as officers labeled it. Police didn't name the boys, aged 17 and 12.

The youngsters said they saw "the thing" twice, once last Friday and once Monday. The boys, who said they saw it from about 50 yards away, described the creature as being covered with hair, walking upright and having human characteristics.

State Police Sgt. Robert Keefe, Roseburg patrol supervisor, said the boys related they didn't tell their parents about it last Friday because "we didn't think anyone would believe us."

They went back Monday to the clearing near an abandoned sawmill where they first saw "the thing." Sure enough, it was there again. The older boy foresightedly had taken along a 30-caliber rifle and fired five shots from less than 50 yards, he told officers.

"It ran off screaming like a cat but louder," the youth said.

The youngsters said they then found humanlike tracks 14 inches long. Police looked, too. The footprints are large, they agreed.

"Could it have been a bear?" Keefe asked. The boys said it couldn't have been; that they'd seen bears before.

Besides, those footprints showed five toes and no claws. Police said they would continue the investigation.

SLICK THINKS BIGFOOT KIN OF ABOMINABLE SNOWMAN

The Humboldt Times, Oct. 15, 1959

A suspected relationship between Bigfoot of Bluff Creek and the Abominable Snowman of the Himalayas has resulted in the formation of the Pacific Northwest Expedition for the purpose of finding out just what kind of a creature Bigfoot really is.

The expedition is sponsored by Tom Slick of San Antonio, Texas, who heads the fourth largest research organization in the work, and S. Kirk Johnson, senior and junior, both also of Texas, along with a group of associates.

Slick and Johnson sponsored the two Slick-Johnson Himalaya expeditions in Nepal in search of the Abominable Snowman, returning from the last expedition in 1957.

In commenting on the expedition, Slick said that there is considerable possibility the creature reported in the Humboldt-Del Norte wilderness could be closely related to that searched for in the Himalayas. In his opinion, if the creature is what it appears to be, its capture could be one of the most important scientific events of all times.

Slick is the founder of the Southwest Research Center, a large philanthropic scientific undertaking in San Antonio, Texas. He points out that the international scientific importance of the project is evident by the fact that there are now four Russian expeditions, two Japanese, two British and two American, all in Asia investigating the Abominable Snowman.

The Northwestern Pacific Expedition includes, in addition to leader Slick, S. Kirk Johnson Jr., associate leader; Robert Titmus, a taxidermist of Anderson, California, deputy leader; about ten other members of this region and from British Columbia, where similar creatures have been reported.

Slick said the scientific work of the expedition will enjoy the advice of a group of scientists working unofficially with Dr. George Agogino of the Department of Anthropology of the University of Wyoming.

The experience of this group engaging in the actual work of the expedition includes more than three years working in the

Himalayas and over 14 months' work by some of the members in the Humboldt-Del Norte area.

It will be the intention of the group to photograph and definitely prove the existence and identification of the creature. Ivan Sanderson, zoologist, and author, is associated with Slick in the expedition.

Slick, besides being involved in the research program, is a noted industrial leader. He is rancher-owner of the Essar (Scientific Research) Ranch in Texas, which has as its main purpose service to mankind in the development of bacon-type hogs, crossing of Brahma cattle with other breeds for hardier types, and similar work.

He is also chairman of the board of the Slick Oil Corporation with headquarters in Houston; head of Slick Secondary Recovery Program at San Antonio; partner in Slick-Moormad Cattle Company; director in Slick Airways, Dresser Industries, Inc., Dallas; Bailey-Selburn Oil and Gas Co., Ltd., Calgary, Canada; and Texas-Mid-Continental Oil and Gas Associated.

Redding Paper, Feb. 27, 1960
Reporter Garth Sanders in Taxidermy Shop of Bob Titmus in Anderson

A girl and her husband dropped into the shop, who formerly lived at Beiber, and gave this account. "She, her mother and a woman friend, were returning from McArthur to Beiber in 1950. As their car topped a rise on the mountain road between Beiber and McArthur, a huge, shadowy form leaped across the highway. Just for an instant the figure was silhouetted against the headlights of the oncoming car. Then it was gone."

The whole thing happened so fast the women were able to conclude that they had seen something fanastic. They spun the car around and returned to McArthur. They thought they had seen some oversized human outcast who was living in the woods. They were under the impression he wore no clothing, or clothing that was unusually tight. They were greeted with laughter when they told their story in McArthur. "This thing may be one of the greatest scientific finds in recent history," Titmus says gravely.

BIGFOOT, IS HE NOW A HUSBAND?

The Humboldt Times, Aug. 17, 1960

WILLOW CREEK—Traveling light and fast both a "Mr. and Mrs. Bigfoot" took the soft dirt of Little Twig's logging spur and Mill Creek road to cross into Mill Creek Canyon Saturday night or early Sunday morning.

The Robert Trenholm family and Mr. and Mrs. Darrell Stewart took both pictures and measurements of the large 15½ inch and 12½ inch tracks. The Trenholms are owners of Little Twig Logging Co. of Willow Creek and their operations are 16 miles from Highway 96 leaving Hoopa Valley at the mouth of Mill Creek.

Little Twig's loader sits on the ridge about 100 yards from Mill Creek road and brush and logs have been cut on the further side of the ridge for another 100 yards. The first track was found to be a deep indentation in the moss as if something had jumped at least eight feet from the top of a log jutting out from the bank. Another and another of the large tracks were joined by the smaller set as they climbed the spur road to the loader. Here the smaller ones went to the left around the rig and the larger ones went determinedly around the right side. They continued on together then for a half mile down Mill Creek itself and turned off down the side of the mountain. Strides of the larger of the tracks were 43 inches average.

A little past midnight Sunday evening trained investigators were at the site for more pictures and plaster casts. At five a.m. Monday morning many loggers coming to work viewed the phenomena with mixed opinions. The tracks could not have appeared on a better traveled thoroughfare than the dusty Mill Creek road, Sunday. A slide between Bluff Creek and Slate Creek on the road construction between Weitchpec and Orleans had brought motorists on a long detour over the Mill Creek road.

EXPERIENCE WITH GIANT IS RECALLED

By BETTY ALLEN, Times Correspondent

WILLOW CREEK—The Byrnes party, consisting of Mr. and Mrs. Peter Byrne and Brian, a brother of Peter, are persistently searching for all available information on the maker of the "Bigfoot" tracks up in the Bluff Creek country. In a recent survey in the Orick area they interviewed Lawrence Omeg who early in the fall of 1958 had reported seeing a giant sized man. Omeg was water tender on the new timber access road and lived in a very small cabin at the beginning of construction.

Omeg explained the incident to the Byrnes and said that about 1:00 a.m. on this moonlight night he heard what seemed to be a knock at the door. Opening it inward a little ways he saw a big man standing right on the tiny porch, not three feet away. To his question of, "What do you want?" the huge thing did not reply but as he looked at it he felt that it wanted something to eat. Beside the door was a small table with food on it and Omeg felt around on it with his hands coming in contact with a candy bar. This he handed to the creature and it took it in both hands and began eating it as it mumbled something which he thought must have been a thanks. Still mumbling, it then turned and went around back of the house and into the creek making strange noises. In the moonlight his feet were bare. Omeg said it looked as though he was wearing a jacket-thing with no sleeves in it. Something loose like a vest. A few days later Omeg thought he saw it again.

Here, having seen, is a man firmly convinced in the reality of "Bigfoot."

$10,000 CASH OFFERED FOR LIVE BIGFOOT

The Humboldt Times

WEAVERVILLE (Trinity Co.)—A $10,000 reward has been offered by the Weaverville Junior Chamber of Commerce for the capture of Bigfoot alive and in good physical condition delivered to the Weaverville Junior Chamber of Commerce on or before 12 noon July 1, 1961. The reward is offered in conjunction with the Bigfoot Celebration and dedication of the Chinese Joss House to be held in Weaverville July 3 and 4.

"We're believers," stated George Killian, general chairman of the celebration. "We believe there is a Bigfoot, and would like to have it, she or he, or whatever it is, brought in to Weaverville."

The $10,000 reward is offered to stimulate interest in the search of the mountainous areas of Trinity and Humboldt counties for the big-footed creature. Tracks have been discovered within four miles of Weaverville; however the many tracks in the western part of Trinity County has led to the establishing of a search party with headquarters in Salyer.

Bigfoot will become the property of the Weaverville Jaycees on his capture and payment of reward. General specifications of Bigfoot established by the Jaycees are: "A creature, man or beast, 7 feet tall, weighing 750 pounds or more with feet measuring 16 inches in length, and covered with hair."

They have eliminated members of the bear or cat family.

The Humboldt Times, July 20, 1963

Bigfoot appears in Hoopa Valley too late for the Fourth of July Celebration and too early for the Big Foot Daze in Willow Creek.

Walking home in late evening a resident of Hoopa Valley, (Peters) saw a strange creature jump out, utter a surprised grunt, jump a five foot fence in one agile leap and disappear into the darkness.

The trail was along Highway 96 just by the fence, and only a short distance past the Old Clover Inn the other side of Jordan's

Store. They ran back to the Oak's Cafe and phoned the Sheriff's Office, but no one was in. Entirely familiar all their lives with animals, they were sure it was not a bear and appeared not too much taller than a man, but of tremendous width.

Earlier in the evening something raided garbage cans at the home of Mr. and Mrs. Francis Francis on the same side of the valley and back against the hill almost even with where the event with Peters took place. The dogs made a fearful din and it left as they opened the door. They did not catch sight of it in the darkness.

ANOTHER 'BIGFOOT' STORY TO ADD TO THE BIG FOOT COLLECTION

Printed in Blue Lake Advocate, Sept. 24, 1964

An abnormal brown bear (some call it "Bigfoot" or the Abominable Snowman") was reported seen at 1:00 a.m., Sunday, September 13th by Benjamin Wilder who told of being awakened while sleeping in his car at his camp on the water pipeline which is at the four thousand foot elevation.

Wilder told how he was awakened by his car being shook up two different times. He said at first he thought it was an earthquake, but thought, with such a hard shake, there should be rocks rolling. Then after the second time, Wilder put on his flashlight and to his surprise there was a large animal standing beside the driver's door of the car.

The animal had its two arms on top of the small car. He said it had long shaggy hair, about three inches long, on the chest. Wilder said he tried to scare it by shouting, but the bear did not move. It only made noises like a hog. Then Wilder honked the car horn. This scared the big animal and it took off, walking on its hind feet, and disappeared over the hill. Wilder said the animal never got down on its four feet, and he never got to see the face of the animal.

This letter indicates the tone several interviewers have come across.

"I was somewhere between Weed and MacDoel, California in

"A Rude Awakening"

June of 1959. I decided to stop for some fresh air,————."

"I was sitting on the fender of my car smoking a cigarette when I heard someone walking down the middle of the road toward me, so I got in the car and turned on the lights to see who it was. I was never so scared in my life. At first I thought it was a big bear walking on its hind legs, but when I put my lights on bright, it looked like a human in a costume, but he was too big. I swear I never saw anything as big in my life. It just stood there a couple of minutes more, then jumped off the road. I was so scared that I just sat there and shook for an hour. I didn't dare move."

"I swore I would never say anything about that night and I couldn't believe that I really saw something so ugly. I was really afraid that I lost my mind."

"I've had a thousand nightmares about that night and I know and swear that I'm going back and hunt that monster down. Believe me, you don't know how lucky you are for not seeing it."

"When I ran across your story about Bigfoot, I knew what it was I had seen."

"God bless you and True Magazine for giving me peace of mind."

(Signed) Anonymous
(Original on file with Mr. I. T. Sanderson)

This incident happened to a young lady from Eureka, California

"When: About nine years ago, at about 10 o'clock in the morning.

"Where: Near the Eel River above Eureka, Calif. At the end of a meadow near the river's edge.

"Under what circumstances: My family and I were fishing on the Eel River. We had been camped in the vicinity for about two weeks and had had poor luck when it came to fishing.

"I used to go for a short walk before breakfast because there was a very pretty meadow about a mile or two from our camp and I used to love to see the mist rise off the grass. I was only about 10 years old at the time and the world of nature was something which both fascinated and enthralled me.

"I entered the meadow and proceeded to cross it in order to reach a small knoll at the other side. When I approached the foot of the knoll, I heard a sound. It was the sound of someone walking, and I thought perhaps my little brother had followed me and was going to jump out and try to scare me.

"I hollered. 'All right, I know you're there.'

"Needless to say, it was not my brother that appeared. Instead, it was a creature that I will never forget as long as I live.

"He stepped out of the bushes, and I froze like a statue. He, or 'it,' was about $7\frac{1}{2}$ or 8 feet tall. He was covered with brown stuff that looked more like a soft down than fur. He had small eyes set close together and had a red look about them. His nose was very large and flat against his face. He had a large mouth with the strangest-looking fangs I have ever seen . . .

"His form was that of a human and he had hands and feet of enormous size, but very human looking.

"However, there was one thing that I have not mentioned, the strangest and most frightening thing of all. He had on clothes! Yes, that's right. They were tattered and torn and barely covered him, but they were still there.

"He made a horrible growling sound that I don't think could be imitated by any living thing. Believe me, I turned and ran as fast as I could. I reached camp winded and stayed scared all the while we were there . . .

"I would be willing to testify to anything I have stated in this letter."

Two doctors of medicine returning from a mass emergency late one night along Route 299 going east from Willow Creek, who said they had nearly run into one, although they had slowed down, thinking it to be somebody signaling for a lift. They said it was at least seven feet tall when it stood up, had straight legs but very long arms, and was clothed in thick lightish brown fur; and who better than (even tired) medical men ought to know? Some of these local stories went back 30 years.

FROM ALBANY, OREG., which is in the Willamette Valley at the foot of Mount Jefferson, comes this brief report published in Fate Magazine's January, 1961, issue:

Albany, Oreg.—The monster of Conser Lake is still on the

loose. The creature reportedly stands on two webbed feet, is 7 to 8 feet high, and with its shaggy white hair somewhat resembles a gorilla. It has kept pace with a truck going 35 m. p. h. Never harmed anyone though.

Further details were contained in a letter to a friend of mine, dated Oct. 27, 1960:

"Creatures (several) last report, being sighted on a farmer's farm. An attempt is being made to contact farmer whom to date wants his name and address held secret. Have made five investigation trips and have for evidence a fingerprint lifted off a house window and a plaster cast of a footprint. Have personal taped accounts of this creature plus many interviews, this includes photographs.

"He is 7' tall, 400 lbs., can move at tremendous speeds, jump tremendous distances. No news items concerning this being have been printed in the Portland papers. He displays extreme cunning, walks and runs erect, appears frustrated, acts as if he would like to communicate. He makes extremely high-pitched sounds. His hair or fur has a slight glow in the dark and is three to four inches long. He walks with feet 19 inches long that make a squeeshy sound. Has been seen in daylight and at night and seen to disappear once into the lake. Will send you complete report as soon as I can.

"Creature first sighted several miles north of Albany, Oreg., in a dense land area approximately three sq. miles. Open land extends all around this area and dotted with farms. Have any ideas how he got there?"

Another remarkable story was told to us at Willow Creek by a gentleman who looked at the plaster cast. He told us that in his high school days on Saturday afternoons, he and his friend usually went hunting or fishing in the surrounding mountains behind his friend's place. On this particular Saturday he had come to pick up his friend at the house, but upon arriving he was surprised to find him sitting in the living room shaking and as white as a sheet. With stuttering voice his friend said he had started to take the garbage out to a little dump behind the shed, and as he rounded the corner of the shed he was startled to see a giant man-type creature with hair all over his body pawing through the rubbish in the dump. He was so petrified that he froze in his tracks. The creature looked up, turned slowly, and ambled back into the forest. The friend threw the garbage down and beat a speedy retreat to the house, too scared even to look out. The gentleman telling this story believed his friend because he had never known him to lie. Besides he had never seen him that scared.

Another eye-witness account comes from south of Willow Creek at the Coastal town of Ft. Bragg, California in 1962. Mr. Jennings who has a ranch there was awakened late at night by his dogs' barking. His brother-in-law who was staying with him at the time, had gone out to see what was the matter and had come back into the house and told them to come quick and see the biggest bear they would ever see. It was standing upright looking over the back fence at the dogs. Mr. Jennings grabbed a gun and a flashlight and stepped out back but didn't see anything.

In the meantime, his brother-in-law stepped out a side door and came face to face with the giant thing he had thought was a bear. Letting out a scream, he stepped backwards, fell down, and went scrambling back into the house on all fours. Mr. Jennings' wife and brother-in-law started to shut the door and got it almost closed when the creature held it open from the outside. Mr. Jennings had come back into the house after hearing the racket from the backyard and told them to step aside and he would shoot the creature. Just then the thing let go of the door and walked upright past the window across the yard and out toward the road. As it walked past the window they could see only the middle of its body as the head and shoulders were above the window and the legs below. They decided to wait until dawn before looking around....

The brother-in-law was so shook up by his experience that it was fifteen or twenty minutes before he could hold a cup of coffee still enough to drink it! Of course, he had come face to face with the creature and the others had not. When questioned about what it looked like, he said it had a flat nose, small, round, black eyes, and a dark rough skin on the face. It had a very bad odor that lingered about the place for some time.

69

Mr. Jennings said it stood about eight feet tall with hair all over its body. He knew it was that tall because his brother-in-law had said it stood a good two feet above the six foot fence in back. They also found a hand print by the door that was quite human except it measured eleven and one-half inches from the palm to the end of the finger! They all agreed it walked upright at all times. They know it was not a bear, wild man, or any kind of ape. What it was for sure, they can only guess.

The foregoing articles and accounts are but a few I have on file; however they are in my opinion the ones most interesting and factual.

In the next chapter we will head further north to Mt. St. Helens and the giant hairy apes (their nickname there). Here we find somewhat the same type and size except these giants seem to have a much meaner temperament. We'll describe a fierce attack on some miners in a canyon (from then on to be called Ape Canyon) during which boulders were hurled down on the miners' cabin as blood curdling screams were heard; screams that sounded all night. We'll read about a skier disappearing, never being seen again although hundreds searched for him. Sound a little terrifying? Well, it is!

"He let out a scream, stepped backward and fell down!"

Chapter 3

GIANT HAIRY APES OF MT. ST. HELENS TERRIFY MINERS AND BAFFLE EVERYONE

For sheer excitement and mystery, the Mt. St. Helens episode rates number one with me.

From the very start it was an odd one. I have lived in Yakima, Washington for 28 years and had never heard of the hairy apes of Mt. St. Helens, although it is only seventy miles from home as the crow flies. I had to go to California to find out about them.

There Betty Allen had a number of newspaper clippings of the startling events that had happened in and around the vicinity of that mountain, mainly since 1924. In that year a group of rugged miners had fled down from the mountains leaving the mine they had been working for three or four years. They claimed to have been attacked by a bunch of huge hairy creatures, barely escaping with their lives.

Betty had wanted more news from that area, so I promised when I returned home I'd go there and look further into the matter.

Prentice Beck, a long time friend and business associate accompanied me on this trip. Prentice is a believer in the subject and as we headed over the mountain pass he remarked in an excited tone "To think we've been this close to something all this time and didn't know it!"

Our first stop was to be at Mr. Ray Wallace's at Toledo (area shown on map 2), Washington. He had been in on "Bigfoot" happenings starting in 1958 in the Bluff Creek area. A few years back he had moved to Toledo and set up business there. Mr. Wallace has changed his mind considerably about who or what caused all the trouble for him while he was building the road up the mountain, in northern California. He no longer thinks someone was trying to pull a trick on him, but is sure now that giant prehistoric-like men were and are in that area.

72

He had a tape recording that was supposed to be a Bigfoot screaming inside a cave. As we listened to it, it didn't sound like what Rod and I had heard in the woods that night we stayed in the cabin. Although we're not sure what it was we had heard. So I cannot say one way or another about the authenticity of Ray's tape.

However, Betty Allen had told us that Mr. Wallace had played the tape to a fellow that had heard one howling on a mountain in the Bluff Creek Area. He was sure it had been one because of the giant human-like tracks found close by the next morning. This gentleman said Ray's tape sounded very close to the same sound he had heard that night! Ray said he was sure that the same kind of creatures that are in northern California existed up around Mt. St. Helens as he had heard a number of stories in recent years coming from loggers and people who had been up there. He also told us of the whereabouts of an old gentleman that claimed to have shot one of the creatures.

The next morning we headed toward the foothills just out of Kelso. We found Mr. Fred Beck living in a small cabin with his son. At first he was reluctant to speak much of his experience; I suppose because of all the ridicule he had taken over the years. But after he had seen that we were sincerely interested in what had happened to him and the tremendous ordeal he had gone through, he soon loosened up and became friendly. There are only two of the old miners left—Mr. Beck and one other who doesn't wish his name used. This is Mr. Beck's astonishing story . . .

Back in about 1921 he and four other prospectors, Gabe Lefever, a Marion Smith, Roy Smith and John Peterson, had started work on a mine in a canyon on the east side of Mt. St. Helens. They had built a solid cabin half way up the canyon wall on a tapered ledge. It was an easy place to get to yet sheltered some from wind and weather. At that time there were no roads in that area and they had to haul their ore out to Kelso 50 or 60 miles by mule.

From time to time they came upon huge suspicious looking tracks which they thought might be those of a big wild renegade Indian. They packed their rifles with them most of the time. Then one day as they were returning to their cabin, they saw a tremendous giant looking out from behind a tree. One of the group quickly fired his rifle, shooting at the thing's head. He was sure he had killed it. However, when they reached the spot where he should have fallen, there was nothing. As they looked up they were surprised to see the giant running over the hill!

The next day they encountered another one close by the edge of the canyon. This time Mr. Beck fired three shots, hitting the creature in the back each time. The giant tumbled over the cliff into the canyon. (Thereafter it has been called Ape Canyon.) They hunted for him but great torrents of water coming off the mountain flooded the canyon floor washing away anything that fell into it. They did not find a sign of him.

As usual they returned to their cabin, had a good supper, and settled down for the night. Then it happened!! Tremendous boulders began pelting their cabin roof followed by loud wailing screams that echoed hideously off the canyon walls. The giants had attacked! They jumped on the roof, rammed their huge bodies against the door and tore at the cabin walls with their hands! If the miners had not built the cabin, roof and all, out of solid ten-inch logs it would never have held together. This fierce attack kept up all night and by dawn there were five scared and weary miners who were thankful to see the light of day. (No doubt this unexpected attack was brought on by the miners themselves when they shot and wounded or killed two of the giants.)

When the miners were sure they had gone they quickly gathered up what was needed to make the trip to town and hurriedly took off down the mountain.

Mr. Beck warned them all to say nothing of their experience as surely no one would believe them. But halfway to town they met two young prospectors and one of the party spilled the beans, telling them of the terrible night before. The two laughed and said they must have had a whiskey party and dreamed the whole thing up. This made Beck furious and he threatened to shoot their heads off if they said another word. The two then went quietly on their way and the miners resumed their trip to town.

Newspapers ran big stories on their being attacked, which some folks believed, while others just laughed and called the whole thing a wild tale. Law enforcement officers and reporters formed

74

"The miners learned this was a mistake."

"The miners barely escaped with their lives!"

a posse and went into the area. They didn't find any hairy ape-like men, although they did find many huge tracks and the miners' cabin torn up badly!

The miners didn't care about the cabin as they never went back, leaving their mine to anyone brave enough to work it.

Mr. Beck seems to be an honest and straightforward person with a keen sense even though he is in his eighties. I believe he told me the truth according to the best of his recollection. After our first meeting I had the privilege of meeting with Mr. Beck many times and he always tried to help in any way he could. Once he let me tape his whole story of the attack.

Another point is that Mr. Beck described the creatures he had soon to closely resemble the ones that had been seen in Canada and California, although he had never heard that such things had been reported anywhere else! We thanked Mr. Beck for all his information and headed up to Mt. St. Helens.

We stopped at Spirit Lake (named so by the Indians because evil spirits were supposed to have lurked there) and soon learned that we came the wrong way to get to Ape Canyon, as the road does not go around the east side of the mountain.

The next morning we stopped at a little logging town by the name of Cougar. We had breakfast at the only cafe in town and asked if anyone had heard lately of the giant hairy apes. The waitress replied smartly, "Yeh, there's one right across the street, he owns the gas station." Then of course the whole place had a good laugh at our expense. More questioning brought more chuckles and wise cracks so I said "Wait a minute, I'll be right back." I went to the pickup and brought back tracks, tape recorder, and hundreds of clippings from the northern California area.

After I showed these and had my say (a few hundred words) you could have heard a pin drop like a falling tree. No more wisecracks or laughing. Instead, slowly, one after another, they would say something like. "Well, I heard this person say such and such about seeing tracks up there, or this one or that one seeing one of the creatures." I bring this out to prove a point.

Many people will try to pass this subject off as a joke, even if they know some valuable information such as these people eventually began to tell us. At first, thinking we were probably just a couple of thrill-seeking tourists, they tried to bluff their way, hoping we would give up and go away. I have run into this situation many times in the past and it usually works out the same. Of course, we must realize that most of these people are good reliable citizens a little confused on this whole matter and not sure of themselves either way. Consequently, they try to pass it off as a good laugh.

Nonetheless, they did come up with some valuable information. They told us of a club called the Ape Cave Explorers. Some of the members lived in the area around Cougar, so we contacted a few of them and found out some startling facts. They told us the caves around Mt. St. Helens are supposed to be the largest lava tubes in the world. There is one such cave that opens at the head of Ape Canyon appropriately called Ape Cave.

At this writing I have not as yet explored any of these caves. However, this summer I am planning an expedition into the area with some device that may attract these giant creatures close enough for a picture. We are planning the same type expedition in the near future into northern California.

As we traveled on down the Lewis River we learned of a fellow named Charley Erion who lives at Woodland, Washington. Charley has a ranch close by the mouth of the Lewis River where it empties into the Columbia River. He related this story to us.

A Portland couple had reported to the local sheriff's office that they had seen a huge creature walking upright along the banks of the Lewis River while they were fishing from a boat. Of course, it scared them badly, as it was only about a hundred feet away. After Charley read the article about the incident in the local newspaper, his first thought was if something had been seen it surely would have left tracks. With this in mind he and his family set out to search the river banks which were nor far from his ranch.

They were well rewarded. They found hundreds of huge human-type tracks measuring a whopping 22 inches long and 10 inches across the ball of the foot. As with most river banks the soil was sandy so the prints showed up perfectly.

Charley also has a logging operation and spends much of his time in the woods so he knows what tracks look like. He said he did not see how those tracks could be faked, as the depth in the sand showed tremendous weight.

He also mentioned he had seen the same type of tracks except smaller up nearer Mt. St. Helens.

I have a number of newspaper clippings from this area that date back as far as 1924. Here are a few of them, plus some eye witness accounts from Central Washington.

Longview Times

'WILD-EYED' RIFLEMAN FLEEING DEVILS OF PEAK LAUNCHED LEGEND OF MT. ST. HELENS APES

(Editor's Note: A durable legend about the mysterious hairy apes of Mt. St. Helens keeps cropping up, as was the case a few months ago when loggers in the area of Ape Canyon waggishly posted hairy-ape road signs. Here, in the first of two articles, a Columbian reporter interviews a man who was there when the legend began.)

By TED VAN ARSDOL
Columbian-Staff Reporter

CHELATCHIE — The July morning in 1924 at the Spirit Lake Ranger Station was beautiful—"like a morning is in the high country," Bill Welch, 68, of Chelatchie, recalled.

What made this morning so different was the man coming from the direction of the ranger station, with a rifle in his hand. Welch could see that the visitor was "pretty wild-eyed" and he recognized him as a man who had a cabin with several others about five or six miles from the station, at the head of what was called Muddy Creek.

"Well, I got 'im," the man said as he slowed down in front of Forest Guard Welch.

"Got who or what ?"

"The mountain devil."

"You mean a cougar?"

"No, the mountain devil."

"You mean a wolverine?"

"No, the mountain devil."

Welch, standing outside the barn eyeing the newcomer warily, recalled him as a man who had been at the station two or three weeks earlier.

Permits were needed to build a fire outside of the campgrounds at Spirit Lake, and this man, in his 50s, had stopped in and asked permission to build a fire in his cabin. Permits were not needed for this.

In the course of the earlier conversation the man told Welch that he and several others had a mine near the cabin. He also volunteered the information that "mountain devils" had been trailing and bothering the men in the last several years. He hadn't seen these "devils," but had viewed some of their tracks.

Welch had thought this was "a little far-fetched"—he didn't think there were any wolves or wolverines in the area.

"It's the first I've heard of it," he had said. "If you run into any of those mountain devils let me know."

"I will," the mountain miner said.

Now, standing facing the gun-carrying mountain man, Welch began wondering if he could get close enough to grab the rifle, and also started worrying about his wife at the ranger station. Had this definitely-disturbed newcomer murdered her?

As the talk continued, Welch was relieved to see his wife come out the door at the ranger station.

The visitor was insisting that he had shot the mountain devil and that it had slid over a bluff at the cabin. He also said that his partners were in a car near the ranger station, after coming down from the cabin with him.

Welch still thought the man had "blown his top" and when he went over to the touring car, which carried three men in front and two in back, he found the others "were just as wild as he was, sitting there clutching their guns."

The man who had walked over to see Welch vowed that they were going back home and would "never come back here again."

The five left shortly afterward, headed for Kelso.

Welch didn't know at the time that he was getting involved in a key incident of the legend of the Mount Saint Helens "hairy apes." The ape label quickly took predominance over the "mountain devils," and the alleged existence of the creatures has been a subject of recurrent speculation over the years.

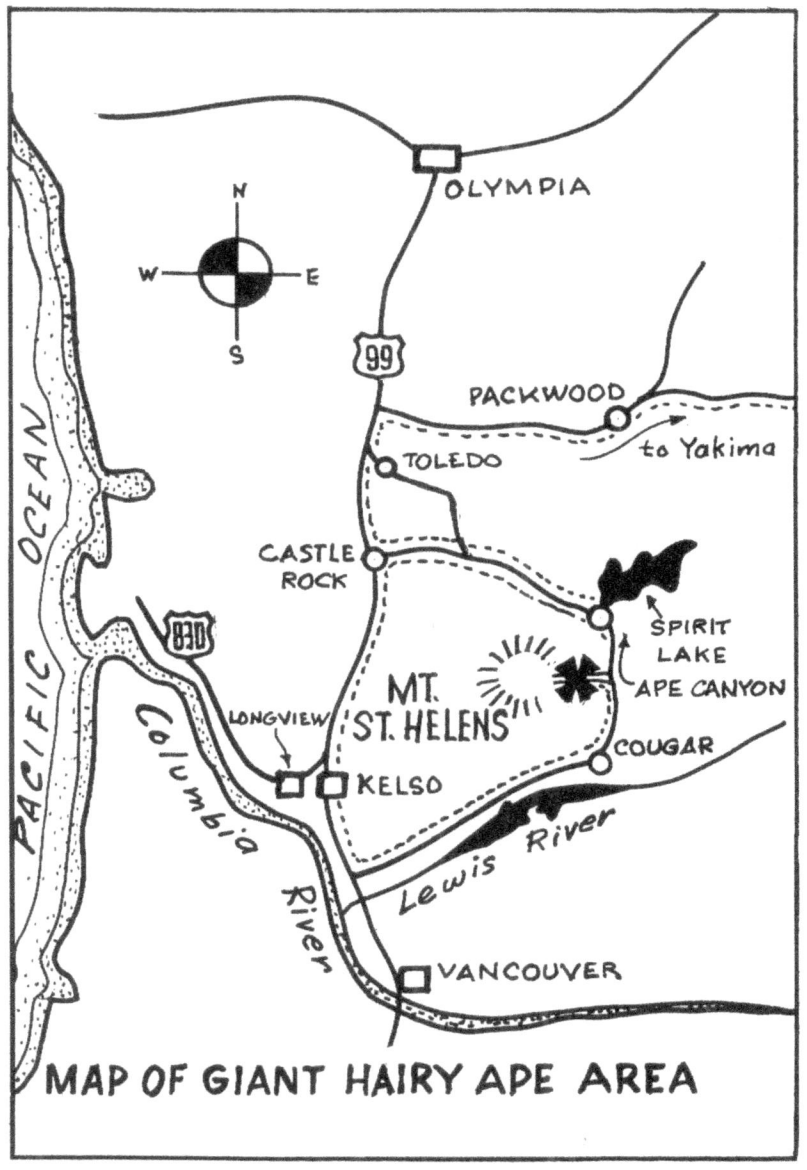

MAP OF GIANT HAIRY APE AREA

Welch talked to his wife, who had been the first person to encounter the departing miners after they left the cabin, and she told him about the man coming to the door, with a rifle across his shoulder, an incident she still remembers vividly today.

The miner's eyes were "glazed" from some evidently shocking experience as he informed Mrs. Welch:

"We got him! We got him!"

"Well—what?"

"The mountain devil."

There was a pause and the miner explained:

"Well, you know your husband told us to let him know if we ever saw any. So I stopped to tell him we saw one and killed it. But we're going out and never coming back."

Mrs. Welch "hardly knew" what he was talking about, but told the gun-toter that her husband was out at the barn, and he walked toward the barn to meet Welch.

After the carload of men had left, Welch called Jim Huffman, the district ranger at Amboy, and told him what had happened.

He still wondered if the miners might have encountered a wolverine. This is a vicious little animal that can wreck cabins, and destroys what it doesn't eat, Welch said. There were rumors of wolverines on Mount Saint Helens about this time, but no one so far as the forest guard knew had seen one.

More details were soon forthcoming from the miners, who were interviewed by a reporter after they had reached Kelso on July 12. "Fight with Big Apes Reported by Miners," was one headline on the reporter's account, which he termed "the strangest story to come from the Cascade Mountains."

The returning miners, Marion Smith, his son Roy Smith, Fred Beck, Gabe Lefever and John Peterson, had encountered "the fabled 'mountain devils' or mountain gorillas" of Mount Saint Helens, according to the reporter, who also stated:

"The men had been prospecting a claim on the Muddy, a branch of the Lewis River about eight miles from Spirit Lake . . .

"They saw four of the huge animals, which are about seven feet tall, weigh about 400 pounds and walk erect. Smith and his companions have seen the tracks of the animals several times in the last six years and Indians have told of the 'mountain devils'

82

for 60 years, but none of the animals ever has been seen before.

"Smith met up with one of the animals and fired at it with a revolver. Thursday Fred Beck shot one, the body falling over a precipice.

"That night the animals bombarded the cabin where the men were stopping with large showers of rocks, many of them large ones, knocking chunks out of the log cabin. Many of the rocks fell through a hole in the roof, and two of the rocks struck Beck, one of them rendering him unconscious for nearly two hours.

"The animals have the appearance of huge gorillas. They are covered with long, black hair. Their ears are about four inches long and stick straight up. They have four toes, short and stubby. The tracks are 13 to 14 inches long.

"These tracks have been seen by forest rangers and prospectors for years.

"The prospectors built a new cabin this year and it is believed it is close to a cave occupied by the animals. Mr. Smith believes he knows the location of the cave."

On the evening that the news was sent by wire from Kelso, Welch recalled, Frank (Slim) Lynch, a Seattle newspaper man, and Burt Hammerstrom, a free-lance writer and brother-in-law of Clarence Darrow, arrived by car at the Spirit Lake Ranger Station.

They had quite a trip in reaching the place, as nine hours were required for a drive from Castle Rock to Spirit Lake in 1924, Welch remembered. He said the road was "not too good," and in some places a driver had to try three or four roads before finding the right one—a side road might meander off into nothing.

Welch later recalled that date as July 14 when the news had been received in Seattle and when he reached the ranger station he and his friend were ready with a lot of questions about the apes who reportedly had rained rocks "as large as a man's head" on the miners' cabin, and as Welch told the story, "had tried to pry the cabin into the Smith Creek abyss."

Huffman, the district ranger, also had arrived at the ranger station and the group planned a trip of inspection to the now-famous cabin.

"Mountain devil" hunters take a rest alongside the cabin where five men were supposed to have been besieged by strange creatures. Bill Welch, now retired and living at Chelatchie, is at upper right in this 1924 photo, standing alongside Burt Hammerstrom, free lance writer. In front, from left, are Frank (Slim) Lynch, Seattle newsman, and Jim Huffman of Amboy, forest ranger for the Spirit Lake district. Photo was loaned by Welch, who was working as a forest guard at Spirit Lake Station in 1924.

LEGENDARY MT. ST. HELENS APEMEN CALLED LEGITIMATE

Longview Times

The legend of the apemen of Mt. St. Helens returns, like hay fever, with summer weather.

The story of the apemen of the beautiful conical mountain situated in the Cascade range of Southwest Washington, is a favorite in the area, but it just may have some basis in fact.

There is more basis to support it than Nepal's Yeti or northern California's "Bigfoot" and probably as much as Loch Ness' monster.

Last summer, two different Portland groups who visited the region reported sighting the monsters, usually described as from 7 to 10 feet tall, hairy and either white or beige colored.

Three persons in a car on a lonely forest road said they saw one of the creatures when it flashed across the headlight beams of their car near the wilderness area which includes such places as "Ape Canyon."

A Portland couple fishing on the Lewis River south of the mountain saw a huge beige figure "bigger than any human" amble off into the brush.

Old timers aren't surprised, just amused. The apeman legend actually is older than the white man's habitation of the Pacific Northwest.

Forestry employees have investigated many reports of the strange creatures. According to Indian legend, the "apes" were the ferocious Selahtik Indians, a band of renegades much like giant apes in appearance who lived like wild animals in the secluded caves of the Cascades.

The first recorded encounter of the apes with white men was in 1924. A group of five prospectors rushed into Kelso to report that a group of great, ape-like creatures had attacked them in the middle of the night.

The miners said they had been working a mine on the east slopes of Mt. St. Helens. During the daytime, they saw some of the apes and fired at them to halt an apparent attack. One of the apes appeared to have been hit and rolled into a deep ravine. That night, according to the account, the apemen hurled rocks onto the cabin and "danced and screamed until daylight."

Then came the "great ape hunt of 1924." Law enforcement officers and a flock of newspapermen made up a posse that went into the area. The armed searchers fired at anything that moved, so the report went. They returned to tell of finding huge footprints, but no apes.

The legend grew from that point for several years, then subsided with only sporadic reports of traces of the apes. Responsible persons, experienced mountaineers and skiers, have given credence to the story.

Bob Lee of Portland, a leader of the 1961 Himalayan expedition and adviser to last year's Himalayan expedition, said last year he had a strange experience. Lee has never claimed to have seen the apes, but said "there was something strange on the high slopes of the mountain."

He was a member of a party that searched for Jim Carter, an experienced skier and mountaineer, who vanished on the mountain in 1950. His disappearance remains a mystery.

At the same time, Lee was a member of the Seattle Mountain Search and Rescue Unit. He described the search for Carter as "the most eerie experience I have ever had." He said that every time he was cut off from the rest of the search party he felt "somebody was watching me."

Carter, he said, had climbed the mountain with some companions on a warm, clear Sunday. He left the group to take a picture and said he would ski to the left of the group. He was never seen again.

His tracks, however, indicate that he suddenly took off down the mountain in a wild, death-defying run that no experienced skier would make—unless he was pursued, Lee said.

The track went in the direction of Ape Canyon. But no trace of Carter or his equipment was found although the area was combed for two weeks. Lee recalled stories of about 25 persons who claim they had encountered the monsters during a 20-year period.

The canyon named for the apes is a lonely, ominous spot in a wild area. It extends to a point near Ape Cave, thought to be the longest unitary lava tube in the world.

There have been many reports of footprints in the area. Some are described as being about 18 inches long and seemingly human.

Unless the creatures are really fuzzy throwbacks, the lost Indian theory seems most likely to some of the fans of the mystery. It has given rise to some suggestions, one of which is to leave well enough alone. The government might take over and shove benefits and subsidies at them—retroactive to the Ice Age.

And that, as well as costing a lot of money, would ruin a very good legend.

Author at one of his camps in Ape Canyon.

APE CANYON HOLDS UNSOLVED MYSTERY

Longview Times
August 1963

By MARGE DAVENPORT, Oregon Journal Staff Writer

SPIRIT LAKE, Wash.—Ape Canyon, the legendary home of the Hairy Apes of Mt. St. Helens apparently swallowed an experienced mountaineer and expert skier in May, 1950.

No trace of Jim Carter, 32, who disappeared from a 20-member climbing party from Seattle was found, although teams of the Northwest's most proficient mountain rescue units combed the area for weeks.

"Carter's complete disappearance is an unsolved mystery to this day," declared Bob Lee, well-known Portland mountaineer who is a member of the exclusive world wide Alpine Club, a leader of the 1961 Himalayan expedition, and adviser to the 1963 American expedition.

Lee said he had never seen one of the monsters, but that there certainly was evidence "that there was something strange on the high slopes of the mountain." He was convinced of this during the search for Carter, he said.

"Dr. Otto Trott, Lee Stark and I finally came to the conclusion that the apes got him," said Lee seriously.

Lee, a member of the Seattle Mountain Search and Rescue unit at the time, describes the hunt for Carter in Ape Canyon as "the most eerie experience I have ever had."

He said that every time he got cut off from the rest of the searchers during the long hunt, he got the feeling that "somebody was watching me."

"I could feel the hair on my neck standing up. It was eerie. I was unarmed, except for my ice ax and, believe me, I never let go of that."

At this point in Lee's story, I could feel my own hair standing up a bit. Ready to shoulder packs for a safari to Ape Canyon to try to determine whether there is any truth to the ape stories, I began to feel a little dubious about the whole expedition.

The rest of Lee's tale about the Seattle man's disappearance didn't do much to reassure me.

It seems that Carter had climbed Mt. St. Helens with a group from Seattle on a warm, clear Sunday. On the way down the mountain, he left the other climbers near a landmark called Dog's Head, at the 8,000-foot level. He told them he would ski around to the left and take a picture of the group as they skied down to timberline.

That was the last time that anyone saw Carter. The next morning searchers found a discarded film box at the point where he had taken a picture.

From here, Carter evidently took off down the mountain in a wild, death-defying dash, "taking chances that no skier of his caliber would take unless something was terribly wrong or he was being pursued," says Lee, who was one of the first searchers to reach Carter's ski tracks.

"He jumped over two or three large crevasses and evidently was going like the devil."

When Carter's tracks reached the precipitous sides of Ape Canyon, the searchers were amazed to see that Carter had been in such a hurry that he went right down the steep canyon walls. But they did not find him at the bottom of the canyon as they expected.

"We combed the canyon, one end to the other for five days. Sometimes there were as many as 75 persons in the search party, but no sign of Carter or his equipment was found," Lee says. After two weeks the search was called off.

Lee, who has lived in the Northwest most of his life, recalls there are about 25 different reports of people attacked by "ape-like men" in the St. Helens and Cascade areas over a 20-year period. One was a group of Boy Scouts from Centralia, he said.

Couldn't we check on that story? As near as he could remember, several of the boys who were taken off the mountain were hysterical after being attacked by the "ape men."

Director Dick Whitney of the regional Boy Scout office in Olympia, Wash., promised to look for a record of the incident. To our surprise he called back to say that he had located the name of the leader and the troop involved in the incident.

"It was a troop under the late Scoutmaster Pease from Centralia," he said. Whitney promised to have Pease's son, who works for the State of Washington, call The Journal as soon as he returns from vacation.

Miners, scouts, Indians, mountaineers and most recently an editor and other reliable Portland residents—the list of persons who have seen the Hairy Apes of Mt. St. Helens is impressive.

Looking west into Ape Canyon. Photo by author.

'MONSTER' SIGHTINGS REKINDLE INTEREST IN MT. ST. HELENS HAIRY APES

By MARGE DAVENPORT, Oregon Journal Staff Writer

Are the legendary Hairy Apes of Mt. St. Helens, which reportedly terrorized early visitors to that area, on the march again?

Strange unidentified monsters reportedly sighted by two different groups of Portland visitors to the Washington area during the past weekend brought knowing nods from old-timers. These "things" have been seen before.

Three persons driving along a remote mountain road east of the Cascade wilderness area early Sunday said they saw a 10-foot, white, hairy figure moving rapidly along the roadside. It was caught in the headlights as their car passed, but they were too frightened to turn around to investigate.

Another Portland woman and her husband fishing on the Lewis River south of Mt. St. Helens saw a huge beige figure, "bigger than any human," along the bank of the river. As they watched, it moved into a thicket with a lumbering gait.

"These reports are shades of the famous Mt. St. Helens apes," according to Forest Ranger Marshall Stenerson, who was stationed at the ranger station for many years. He has listened to and investigated many reports about the strange monsters that supposedly inhabit the slopes and the remote, wild country around the beautiful mountain.

Stenerson is now stationed in Portland, but while he was in charge of the mountain ranger station he instigated an investigation into the history and legends of the St. Helens area. This investigation revealed that the stories of the apes on Mt. St. Helens are older than the white man's inhabitation of the Northwest.

The Clallam Indian tribe claims the apes are the ferocious Selahtik Indians, a tribe of renegade marauders much like giant apes in appearance, who lived like animals in the caves in the high Cascades.

Evidently the white man's first encounter with the apes was a wild one. In 1924 Marion Smith and five miners rushed into Kelso, Wash., to report that a band of great ape-like creatures had attacked them in the middle of the night.

Smith said they had been working in a mine on the east side of Mt. St. Helens. They encountered some of the ape-men on the mountainside during the daytime and fired on them to halt an attack at that time. One of the huge creatures was believed slain, and the body rolled over a cliff into a deep ravine, destined, thereafter, to be known as Ape Canyon.

The attack continued after dark, Smith told the Cowlitz County Sheriff, and the apemen pelted their cabin all night with rocks, and danced and screamed until daylight.

They described the mountain "devils" as being at least 7 feet tall and covered with long, black hair. Their arms were long and trailed, they said.

The "great ape hunt of 1924" followed. The sheriff led a large party out of Kelso on an eerie trip to Mt. St. Helens, with all participants armed. They found huge footprints around the miners' cabin, but never saw an ape. Nevertheless, the miners never went back to their mine.

Inspired by this "white man legend," an employee at the ranger station later had a lot of fun with a large foot form. From time to time he left its imprints on the lake shore. This caused a lot of excitement, and later, when someone discovered all tracks were of the same right foot, he admitted the hoax.

However, the ape legend has persisted and more fuel has been added to the fire from time to time as intermittent reports have come in about persons sighting strange figures on the mountain sides, or hearing weird noises in the wilderness.

However, the sightings last weekend were the first reported for several years. Are the old-timers right when they surmise that the hairy ape-men may be on the move again?

GIANT MYSTERY TRACKS FOUND ALONG LOWER LEWIS RIVER BANKS

HAS HAIRY APE WANDERED FAR FROM HOME?
By MARGE DAVENPORT, Journal Staff Writer
August 1963

WOODLAND, Wash. — Is one of the Hairy Apes of Mt. St. Helens lurking in the thickets along the lower Lewis River?

The folks who live down there think so, and they all are looking over their shoulders whenever they go outside these days.

The tracks of a "monster" or whatever, found on the riverbank Sunday by 11-year-old Jimmie Erion, his brother and his father, Charlie Erion, really convinced them.

The tracks are 16 to 18 inches long and 8 inches wide. They show that whatever made them took 4 to 6-foot strides. Residents inspecting the tracks estimate that "the thing" weighed about 1,000 pounds.

A few days before the monster footprints were found, a Portland couple fishing in a boat on the river had seen a lumbering 10-foot-tall beige-colored monster in the brush along the river. They said it lumbered off into a thicket as they watched.

Erion, a logger who owns a 320-acre cattle and truck-garden ranch on the Lewis River, was walking with his sons Sunday when they discovered the tracks. They quickly summoned Arland Brawner, Portland businessman, visiting in the area, and a boom man who operates the log dump nearby.

The boom man, who took pictures of the tracks, asked that his name not be used because he was afraid he would be kidded about "seeing things." He said that he was sure the prints were not a fake.

"It would have been impossible for anyone to have gotten to the place to make the tracks unless he did it with a helicopter," he declared.

The tide is about 4 feet in the Lewis River where the tracks were found. The prints were visible clearly in the sand and mud below the tide line. There were no human tracks or signs of a boat near them, the boom man pointed out. The tracks circled out of the river, onto a road and then led back to the water.

The boom man said he had no idea what made the tracks except that "they are human, only way too big."

The area surrounding the place where the prints were found is a tangle of brush for miles and would make an ideal hiding place for anything that wanted to stay in the wilds, he said.

The following is an excerpt from a tape recorded interview with Paul Manly, Business Editor of the Oregon Journal. This interview was made by Trippett at the offices of the Oregon Journal Publishing Co. in Portland, Oregon on the afternoon of Sunday, October 20, 1963.

Manly—

I've lived in Cleveland all my life. I had only been up through this area once before myself. The area from Goldendale to Toppenish is very uninhabited. There isn't a town on the road and as a matter of fact, there's only one gas station out of the pine tree country into the sagebrush country. There are no trees at all, just sagebrush.

We drove the Indian girls up to Toppenish on Saturday night, we must have left them about 11:00 o'clock. We stopped off for a bite to eat before heading back to Goldendale. We had remarked on the way up what a beautiful clear night it was. The moon was full and there wasn't a cloud in the sky.

We were coming back from Toppenish, heading southeast. As you come up out of Toppenish you start up through the pass. After traveling a bit we came upon a car pulling another, so I pulled out to pass. Having to pass two car lengths, I was paying a little more attention to the highway than I normally would have. Just as I pulled along side of him, on the left hand shoulder of the road, this shape appeared right at the shoulder of the road. My personal reaction was, "What are they doing leaving a big tree stump on the shoulder of the road where somebody is going to run into it." It was that close and I got the impression that it was a tree stump that had been lopped maybe nine or ten feet up, struck off by the wind, lightning, or something like that. You could judge the height fairly well, because of the position of it in relation to the car. So, this was my first reaction. Then I thought to myself, "This is sagebrush country." It kind of made the hair stand up on the back of my head. I pulled in front of the other car and I didn't say anything to the two girls who were still with me. I wasn't going to say a word to them. I thought I had been seeing things. Just as we closed with the other car one of the girls said to me, or the world at large, "Did you see that?" She had started to turn around to look at whatever it was she saw and in so doing she faced the other girl. We were all in the front seat, and she caught this expression of terror on her face, although she wasn't saying a word.

I realized all three of us must be seeing something. They had seen it before I had. They said they saw it come up out of this low sloping ditch at the side of the road. It came up the bank in great strides and they thought it was going to run in front of the car when it stopped dead on the shoulder of the road. We agreed that it was about nine feet tall and light grey or beige in color.

When the beautician got back to work in Vancouver the following week, she told the two Indian girls about our having seen some unexplained object. The Indian girl's reaction was, "Oh, you saw a Stick Indian." According to their folklore there is a tribe of renegade Indians whom they prefer to believe are an offshoot of the Klickitats. According to these girls' description they are about eight or nine feet tall and shaggy in appearance, due to clothing or long hair. Another characteristic is a terrific stench.

We were intrigued enough that we went back the following weekend, to establish in daylight the approximate location of what we saw. We established that it was about a mile from where a creek comes down from the east and flows into Satus Creek. There is nothing but sagebrush in that area. We did not find any tracks. This was on Saturday the 23rd. That would mean we saw "it" July 17 between 1:00 and 1:30 in the morning.

Chapter 4

CANADA SASQUATCH INDIAN LEGENDS COME TO LIFE

Canada and Alaska are the last frontiers in North America, but you may not realize, as I didn't, just how much a frontier they really are. For instance, in British Columbia, there is one section that covers 250,000 square miles of trackless wilderness. This section has only a single main road through it and a few small towns and villages.

It is from that section that many strange stories have evolved over the years. One of these stories came out in the following article that appeared in True Magazine . . .

A NEW LOOK AT AMERICA'S MYSTERY GIANT

By IVAN T. SANDERSON
Copyright 1960, Fawcett Publications, Inc.

On a sunny October day in 1955 a young man named William Roe decided to take a day off from his work on a road-building crew and go hunting. What he did on that day, and most particularly what he saw, electrified everyone who heard of it. For Roe came face-to-face with one of the huge, hairy human-like creatures which Americans know as *Bigfoot* and which Canadians call the *Sasquatch*.

Stories about Canada's version of the Abominable Snowman are almost as old as the country itself, but Roe's account was so detailed and convincing that it could not be laughed off by those cynics who cannot accept anything they do not understand. And from the day it became known, the Sasquatch began emerging from the misty land of legend into the cold light of the twentieth century.

Roe's account of his remarkable experience is a matter of public record. He has described it in his own words, and has made a sworn statement as to its authenticity before a public solicitor. Before letting him tell his story, there are two things I would like to make clear. First, Roe is a man who has spent most of his life in the outdoors, he is a veteran hunter, and when he sees a bear he does not get hysterical and think it is something else.

Second, while "sworn statements" may not cut too much ice in this country, they mean a very great deal in Canada and other parts of the British Empire. Canadians have an intense respect for the law, and their laws are quite a lot more stringent than ours. If you make a sworn statement to legal authority in the presence of witnesses you sign your honor to it. If you lie, you are held responsible. If it is proved for any reason later that you lied, you have committed perjury, and you are liable for whatever injuries your lies may have caused. A Canadian thinks more than twice before he goes before a justice of the peace and makes a sworn statement. So, with the kind permission of Mr. Roe himself and of Mr. John Green of the *Agassiz-Harrison Advance,* who persuaded Roe to make his experience known, I give you the former's statement verbatim. It reads:

Affidavit

I, W. Roe, of the City of Edmonton, in the Province of Alberta, make oath and say,

(1) That the exhibit A. attached to this, my affidavit, is absolutely true and correct in all details.

Sworn before me in the City of Edmonton, Province of Alberta, this 26th day of August, A.D. 1957.

(Signed) William Roe
(Signed) by W. H. Clark
Assistant Claims Agent
Number D. D. 2822

"EXHIBIT A.

"Ever since I was a small boy back in the forests of Michigan I have studied the lives and habits of wild animals. Later, when I supported my family in northern Alberta by hunting and trapping, I spent many hours just observing the wild things. They fascinated me. The most incredible experience I ever had with a wild creature occurred near a little place called Tete Jaune Cache, British Columbia, about 80 miles west of Jasper, Alberta.

"I had been working on the highway near this place Tete Jaune Cache for about two years. In October, 1955, I decided to climb five miles up Mica Mountain to an old deserted mine, just for something to do. I came in sight of the mine about 3 o'clock in the afternoon after an easy climb. I had just come out of a patch of low brush into a clearing, when I saw what I thought was a grizzly bear in the brush on the other side. I had shot a grizzly near that spot the year before. This was only about 75 yards away, but I didn't want to shoot it, for I had no way of getting it out. So I sat down on a small rock and watched, with my rifle in my hand.

"I could just see part of the animal's head and the top of one shoulder. A moment later it raised up and stepped out into the opening. Then I saw it wasn't a bear.

"This, to the best of my recollection is what the creature looked like and how it acted as it came across the clearing directly towards me. My first impression was of a huge man about six feet tall, almost three feet wide, and probably weighing somewhere near 300 pounds. It was covered from head to foot with dark brown, silver-tipped hair. But as it came closer I saw by its breasts that it was a female.

"And yet, its torso was not curved like a female's. Its broad frame was straight from shoulder to hip. Its arms were much thicker than a man's arms and longer, reaching almost to its knees. Its feet were broader proportionately than a man's, about five inches wide in the front and tapering to much thinner heels. When it walked it placed the heel of its foot down first, and I could see the grey-brown skin or hide on the soles of its feet.

"It came to the edge of the bush I was hiding in, within twenty feet of me, and squatted down on its haunches. Reaching out its hands it pulled the branches of bushes towards it and stripped the leaves with its teeth. Its lips curled flexibly around the leaves as it ate. I was close enough to see that its teeth were white and even.

"The shape of this creature's head somewhat resembled a Negro's. The head was higher at the back than at the front. The nose was broad and flat. The lips and chin protruded farther than its nose. But the hair that covered it, leaving bare only the parts of its face around the mouth, nose and ears, made it resemble an animal as much as a human. None of this hair, even on the back

98

of its head, was longer than an inch, and that on its face much shorter. Its ears were shaped like a human's ears. But its eyes were small and black like a bear's. And its neck also was unhuman, thicker and shorter than any man's I have ever seen.

"As I watched this creature I wondered if some movie company was making a film in this place and that what I saw was an actor made up to look partly human, partly animal. But as I observed it more I decided it would be impossible to fake such a specimen. Anyway, I learned later there was no such company near that area. Nor, in fact, did anyone live up Mica Mountain, according to the people who lived in Tete Jaune Cache.

"Finally, the wild thing must have gotten my scent, for it looked directly at me through an opening in the brush. A look of amazement crossed its face. It looked so comical at that moment I had to grin. Still in a crouched position, it backed up three or four short steps, then straightened up to its full height and started to walk rapidly back the way it had come. For a moment it watched me over its shoulder as it went, not exactly afraid, but as though it wanted no contact with anything strange.

"The thought came to me that if I shot it I would possibly have a specimen of great interest to scientists the world over. I had heard stories about the *Sasquatch,* the giant hairy Indians that live in the legend of the Indians of British Columbia and also, many claim, are still in fact alive today. Maybe this was a *Sasquatch,* I told myself.

"I levelled my rifle. The creature was still walking rapidly away, again turning its head to look in my direction. I lowered the rifle. Although I have called the creature 'it,' I felt now that it was a human being, and I knew I would never forgive myself if I killed it.

"Just as it came to the other patch of brush it threw its head back and made a peculiar noise that seemed to be half laugh and half language, and which I could only describe as a kind of a whinny. Then it walked from the small brush into a stand of lodge-pole pines.

"I stepped out into the opening and looked across a small ridge just beyond the pines to see if I could see it again. It came out on the ridge a couple of hundred yards away from me, tipped its head back again, and again emitted the only sound I had heard it make,

"I levelled my rifle. Then I realized that the creature was a human being . . ."

but what this half-laugh, half-language was meant to convey I do not know. It disappeared then, and I never saw it again.

"I wanted to find out if it lived on vegetation entirely or ate meat as well, so I went down and looked for signs. I found it in five different places, and although I examined it thoroughly, could find no hair or shells, or bugs or insects. So I believe it was strictly a vegetarian.

[Author's note: I presume he is referring here to droppings or faeces of this creature, of which he says he found evidence in five different places.]

"I found one place where it had slept for a couple of nights under a tree. Now, the nights were cool up the mountain at this time of year especially, and yet it had not used a fire. I found no signs that it possessed even the simplest of tools. Nor did I find any signs that it had a single companion while in this place.

"Whether this creature was a *Sasquatch* I do not know. It will always remain a mystery to me unless another one is found.

"I hereby declare the above statement to be in every part true, to the best of my powers of observation and recollection."

(Signed) William Roe

Stories about the *Sasquatch* have been appearing in print from time to time since the 1860's, and I have clippings in my files from almost every year since the early 1920's. But the modern history of the *Sasquatch* really dates from September, 1941, when one of these creatures paid a visit—in broad daylight—to an Indian family named Chapman. While the Amerindian stories have usually been dismissed as legend, or laughed off because Indians are not supposed to be reliable, this experience was accompanied by too much physical evidence to be ignored.

The Chapman family consisted of George and Jeannie Chapman and children numbering, as of my visit, four. Mr. Chapman worked on the railroad, and was living at that time in a small place called Ruby Creek, 30 miles up the Fraser River from Agassiz, British Columbia, in Canada's great western province.

It was about 3 in the afternoon of a sunny, cloudless day when Jeannie Chapman's eldest son, then aged 9, came running to the house saying that there was a cow coming down out of the woods at the foot of the nearby mountain. The other kids, a boy aged 7

101

and a little girl of 5, were still playing in a field behind the house bordering on the railroad track.

Mrs. Chapman went out to look, since the boy seemed oddly disturbed, and then saw what she at first thought was a very big bear moving about among the bushes bordering the field beyond the railroad tracks. She called the two smaller children who came running immediately. Then the creature moved out onto the tracks and she saw to her horror that it was a gigantic man covered with *hair,* not fur. The hair seemed to be about four inches long all over, and of a pale yellow-brown color. To pin down this color Mrs. Chapman pointed out to me a sheet of lightly varnished plywood in the room where we were sitting. This was of a brownish-ochre color.

This creature advanced directly toward the house and Mrs. Chapman had, as she put it, "much too much time to look at it" because she stood her ground outside while the eldest boy—on her instructions—got a blanket from the house and rounded up the other children. The kids were in a near panic, she told us, and it took two or three minutes to get the blanket, during which time the creature had reached the near corner of the field only about 100 feet away from her. Mrs. Chapman then spread the blanket and, holding it aloft so that the kids could not see the creature or it them, she backed off on the double to the old field and down on to the river beach out of sight, and then ran with the kids downstream to the village.

I asked her a leading question about the blanket. Had her purpose in using it been to prevent her kids seeing the creature, in accord with an alleged Amerindian belief that to do so brings bad luck and often death? Her reply was both prompt and surprising. She said that, although she had heard white men tell of that belief, she had not heard it from her parents or any other of her people whose advice regarding the so-called *Sasquatch* had been simply not to go farther than certain points up certain valleys, to run if she saw one, but not to struggle if one caught her as it might squeeze her to death by mistake.

"No," she said, "I used the blanket because I thought it was after one of the kids and so might go into the house to look for them instead of following me." This seems to have been sound

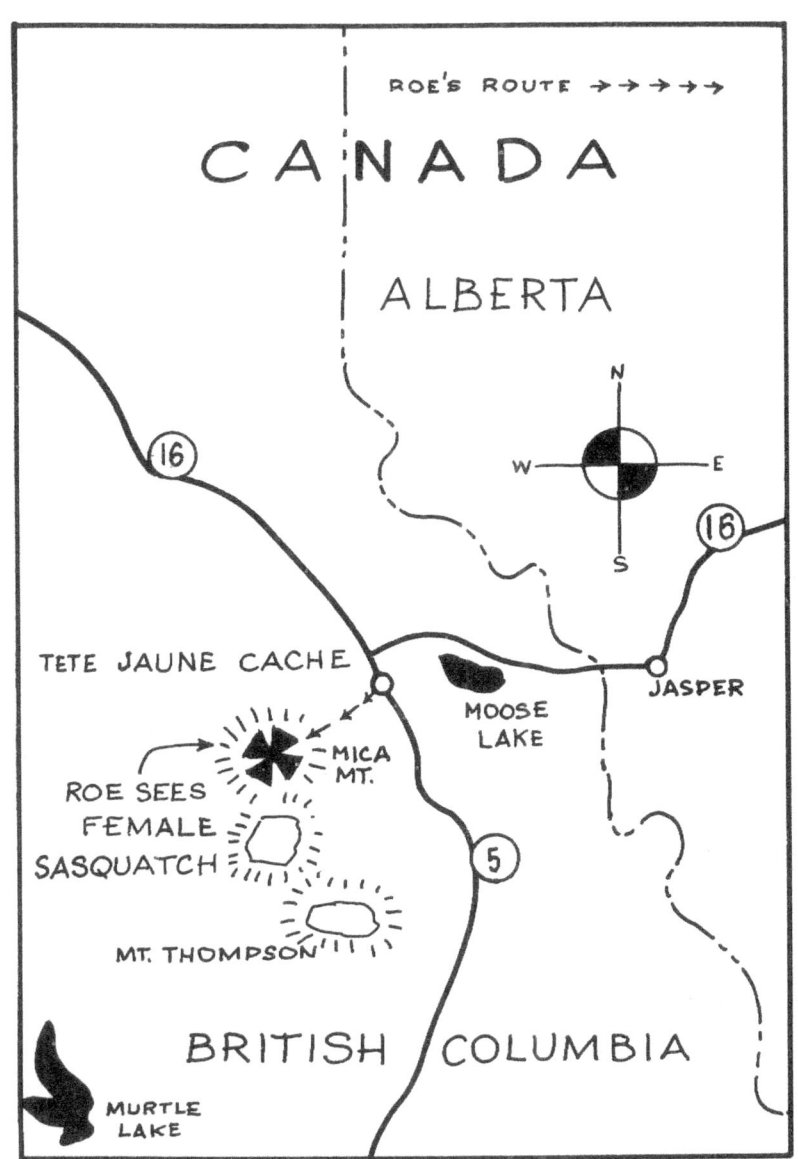

ROE'S ROUTE → → → → →

CANADA

ALBERTA

N
W E
S

16

16

TETE JAUNE CACHE

JASPER

MOOSE
LAKE

MICA
MT.

ROE SEES
FEMALE
SASQUATCH

5

MT. THOMPSON

BRITISH COLUMBIA

MURTLE
LAKE

logic as the creature did go into the house and also rummaged through an outhouse pretty thoroughly, hauling from it a 55 gallon barrel of salt fish, breaking this open, and scattering its contents about outside. (The irony of it is that all those three children did die within three years; the two boys by drowning, and the little girl on a sickbed. And just after I interviewed the Chapmans they also were drowned in the Fraser River when a rowboat capsized.)

Mrs. Chapman told me that the creature was about 7½ feet tall. She could easily estimate the height by the various fence and line posts standing about the field. It had a rather small head and a very short, thick neck; in fact really no neck at all, a point that was emphasized by William Roe and by all others who claim to have seen one of these creatures. Its body was entirely human in shape except that it was immensely thick through its chest and its arms were exceptionally long. She did not see the feet which were in the grass. Its shoulders were very wide and it had no breasts, from which Mrs. Chapman assumed it was a male, though she also did not see any male genitalia due to the long hair covering its groin. She was most definite on one point: the naked parts of its face and its hands were much darker than its hair, and appeared to be al-most black.

George Chapman returned home from his work on the railroad that day shortly before 6 in the evening and by a route that by-passed the village so that he saw no one to tell him what had happened. When he reached his house he immediately saw the wood-shed door batttered in, and spotted enormous humanoid footprints all over the place. Greatly alarmed — for he, like all his people, had heard since childhood about the "big wild men of the mountains," though he did not hear the word *Sasquatch* till after this incident—he called for his family and then dashed through the house. Then he spotted the foot tracks of his wife and kids going off toward the river. He followed these until he picked them up on the sand beside the river and saw them going off downstream without giant ones following.

Somewhat relieved, he was retracing his steps when he stumbled across the giant's foot-tracks on the river bank farther upstream. The giant had come out of the potato patch, which lay between the house and the river, had milled about by the river, and then gone back through the old field toward the foot of the mountains.

Returning to the house, relieved to know that the tracks of all four of his family had gone off downstream to the village, George Chapman went to examine the woodshed. In our interview, after 18 years, he still expressed voluble astonishment that any living thing, even a 7-foot-6-inch man with a barrel chest could lift a 55-gallon tub of fish out of the narrow door of the shack and break it open without using a tool. He confirmed the creature's height after finding a number of long brown hairs stuck in the slabwood lintel of the doorway.

George Chapman then went off to the village to look for his family, and found them in a state of calm collapse. He gathered them up and invited his father-in-law and two others to return with him, for protection of his family when he was away at work.

The foot-tracks returned every night for a week and on two occasions the dogs that the Chapmans had taken with them set up the most awful racket at exactly 2 o'clock in the morning. The *Sasquatch* did not, however, molest them or, apparently, touch either the house or the woodshed. But the whole business was too unnerving and the family finally moved out. They never went back.

After a long chat about this and other matters, Mrs. Chapman suddenly told us something very significant just as we were leaving. She said, "It made an awful funny noise." I asked her if she could imitate this noise for me but it was her husband who did so, saying that he had heard it at night twice during the week after the first incident. He then proceeded to utter exactly the same strange, gurgling whistle that the men in California, who said they had heard *Bigfoot* call, had given us. This is a sound I cannot reproduce in print, but I can assure you that it is unlike anything I have ever heard given by man or beast anywhere in the world.

To me, this information is of the greatest significance. That an Amerindian couple in British Columbia should give out with exactly the same strange sound in connection with a *Sasquatch* that two highly educated white men did, over 600 miles south in connection with California's *Bigfoot,* is incredible. If this is all a hoax or a publicity stunt, or mass-hallucination, as some people have claimed, how does it happen that this noise—which defies

105

description—always sounds the same no matter who has tried to reproduce it for me?

These were probably the last words on the *Sasquatch* that the Chapmans uttered and I absolutely refuse to listen to anyone who might say that they were lying. Admittedly, honest men are such a rarity as possibly to be non-existent, but I have met a few who could qualify and I put the Chapmans near the head of the list.

What on earth had they to gain by making up such a story? All they had ever gotten in return for doing so in the first place was ridicule and insults to their ancient race. And we just walked up to them unannounced on a railroad track and they did not tell us what we "wanted to know," because we never said exactly what that was.

And, besides, there were plenty of white men who went and looked at those tracks at that time, and they weren't all in cahoots and involved in some devilish plot to defraud the public.

The experience of the Chapman family kicked the lid off a fairly large pot that had been brewing for a long time.

A Mr. John W. Burns, now of San Francisco, had for many years been collecting every scrap of information on this subject and had published a number of articles on it. Actually, it was he who had bestowed the name *Sasquatch* on what the Amerindians had previously called, in their various languages and dialects, merely "Wild Men of the Mountains." Mr. Burns was a schoolteacher and had been an Indian agent. He is a man of much erudition.

There was a long and rather full tradition about the *Sasquatch* in British Columbia, and especially on Vancouver Island, where so many sightings have been reported. Vancouver Island is enormous. It is very rugged, clothed in the densest forest, and is, even today, for the most part unexplored. What is more, it was the first part of the Northwest Pacific Rain-Forest to be invaded by roads, and thus, first of these unexplored regions where sightings could have been made.

Getting back to the various accounts, I would like to emphasize again that they show a remarkable continuity and similarity that goes beyond the possibilities of coincidence. And you must bear in mind that the widely assorted people who saw a *Sasquatch* did not

106

know what had been reported before; in fact, a great many of them were completely unaware that any such thing had even been seen anywhere in the world.

Why and how should responsible, sensible men like William Roe make up all these details, details which so exactly coincide with little incidental items recorded by Sherpas in Nepal, bulldozer operators in California, Amerindians on Vancouver Island, teen-agers going home from a dance in Agassiz, and so forth? What, I ask the skeptics, is the idea? Is there some sort of international plot and, if so, why do the plotters persist in getting unknown people in obscure places to give out incredible statements?

Let me close with one final Sasquatch sighting, as this was the one which first made news throughout the world. It happened in 1956 when a Mr. Stanley Hunt of Vernon, British Columbia—a man who had not previously been in any way interested in this matter, nor, in fact, had even heard of it outside of some joking references in local newspapers—was driving through the small township of Flood on the Fraser River. Shortly after dark he saw a large humanoid clothed in "grey hair" cross the road while another similar creature "gangly, not stocky like a bear, stood in the bush beside the road." Flood is immediately adjacent to Ruby Creek. So we are right back where we started.

The matter of Bigfoot in California is, at the moment of writing, a very live issue, and several people are putting a good deal of money into an extensive investigation. But Sasquatch is no less important. This creature has been told about by the Amerindians for centuries, and allegedly seen by white men for more than a century, and it is still being encountered today. Are we just going to let this thing slip through our fingers by sitting back and laughing it off?

Here is something profoundly alive in our very midst that certainly needs proper and intelligent study, and some serious effort expended upon it. And it is a matter that might produce one of the greatest scientific discoveries of our time.

In the months that immediately followed William Roe's account of his encounter with Sasquatch, he received a number of letters from interested people asking questions about the episode. One letter, however, didn't ask a question . . . it answered a few that Roe, himself, had had.

107

The letter came from a Mr. Albert Ostman. While Mr. Roe had been undecided as to whether or not the creature he had seen was human, Mr. Ostman was not! Mr. Ostman wrote that he, himself, had been kidnapped by one of these huge creatures and they definitely were human. Here is his incredible story: . . .

AGASSIZ-HARRISON

The Advance

Friday, May 5, 1961

KIDNAPPED PROSPECTOR LIVED SEVERAL DAYS WITH GIANT FAMILY

I have always followed logging and construction work. This time I had worked over one year on a construction job, and thought a good vacation was in order. B.C. is famous for lost gold mines. One is supposed to be at the head of Toba Inlet—why not look for this mine and have a vacation at the same time? I took the Union Steamship boat to Lund, B.C. From there I hired an old Indian to take me to the head of Toba Inlet.

This old Indian was a very talkative old gentleman. He told me stories about gold brought out by a white man from this lost mine. This white man was a heavy drinker—spent his money freely in saloons. But he had no trouble in getting more money. He would be away a few days, then come back with a bag of gold. But one time he went to his mine and never came back. Some people said a Sasquatch had killed him.

At that time I had never heard of Sasquatch. So I asked what kind of an animal he called a Sasquatch? The Indian said:

"They have hair all over their bodies, but they are not animals. They are people. Big people living in the mountains. My uncle saw the tracks of one that were two feet long. One old Indian saw one over eight feet tall."

I told the Indian I didn't believe in their old fables about mountain giants. It might have been some thousands of years ago, but not nowadays.

The Indian said: "There may not be many, but they still exist."

108

Late one day I found an exceptionally good campsite. It was two good sized cypress trees growing close together and near a rock wall with a nice spring just below these trees. I intended to make this my permanent camp. I cut lots of brush for my bed between these trees. I rigged up a pole from this rock wall to hang my pack sack on, and I arranged some flat rocks for my fireplace for cooking. I had a really classy setup. I shot a grouse just before I came to this place. Too late to roast that tonight—I would do that tomorrow.

And that is when things began to happen.

I am a heavy sleeper, not much disturbs me after I go to sleep, especially on a good bed like I had now.

Next morning I noticed things had been disturbed during the night. But nothing missing that I could see. I roasted my grouse on a stick for my breakfast—about 9:00 a.m. I started out prospecting.

That night I filled up the magazine of my rifle. I still had one full box of 20 shells in my pack, besides a full magazine and six shells in my coat pocket. That night I laid my rifle under the edge of my sleeping bag. I thought a porcupine had visited me the night before and porkies like leather, so I put my shoes in the bottom of my sleeping bag.

Next morning my pack sack had been emptied out. Some one had turned the sack upside down. It was still hanging on the pole from the shoulder straps as I had hung it up. Then I noticed one half-pound package of prunes was missing. Also my pancake flour was missing, but my salt bag was not touched. Porkies always look for salt so I decided it must be something other than porkies. I looked for tracks but found none. I did not think it was a bear, they always tear up and make a mess of things. I kept close to camp these days in case this visitor would come back.

I climbed up on a big rock where I had a good view of the camp, but nothing showed up. I was hoping it would be a porky, so I would get a good porky stew. These visits had now been going on for three nights.

I intended to make a new campsite the following day, but I hated to leave this place. I had fixed it up so nice, and these two cypress trees were bushy. It would have to be a heavy rain before

"A long trip . . . in a sleeping bag."

I would get wet, and I had good spring water and that is hard to find.

This night it was cloudy and looked like it might rain. I took special notice of how everything was arranged. I closed my pack sack, I did not undress, I only took off my shoes, put them in the bottom of my sleeping bag. I drove my prospecting pick into one of the cypress trees, so I could reach it from my bed. I also put the rifle alongside me, inside my sleeping bag. I fully intended to stay awake all night to find out who my visitor was, but I must have fallen asleep.

I was awakened by something picking me up. I was half asleep and at first I did not remember where I was. As I began to get my wits together, I remembered I was on this prospecting trip, and in my sleeping bag.

My first thought was—it must be a snow slide, but, there was no snow around my camp. Then it felt like I was tossed on horseback, but I could feel whoever it was, was walking.

I tried to reason out what kind of animal this could be. I tried to get out my sheath knife and cut my way out, but I was in almost a sitting position, and the knife was under me. I could not get hold of it, but the rifle was in front of me. I had a good hold of that, and had no intention of letting go. At times I could feel my pack sack touching me, and could feel the cans in the sack touching my back.

After what seemed like an hour, I could feel we were going up a steep hill. I could feel myself rise for every step. What was carrying me was breathing hard and sometimes gave a slight cough. Now I knew this must be one of the mountain Sasquatch giants the Indians told me about.

I was in a very uncomfortable position—unable to move. I was sitting on my feet, and one of my boots in the bottom of the bag was crossways with the hobnail sole up across my foot. It hurt me terribly, but I could not move.

It was very hot inside. It was lucky for me this fellow's hand was not big enough to close up the whole bag when he picked me up—there was a small opening at the top, otherwise I would have choked to death.

Now he was going downhill. I could feel myself touching the

ground at times and at one time he dragged me behind him and I could feel he was below me. Then he seemed to get on level ground and was going at a trot for a long time. By this time, I had cramps in my legs—the pain was terrible. I was wishing he would get to his destination soon. I could not stand this type of transportation much longer.

Now he was going uphill again. It did not hurt me so bad. I tried to estimate the distance and directions. As near as I could guess we were about three hours traveling. I had no idea when he started as I was asleep when he picked me up.

Finally, he stopped and let me down. Then he dropped my pack sack. I could hear the cans rattle. Then I heard chatter—some kind of talk I did not understand. The ground was sloping so when he let go of my sleeping bag, I rolled over head first down hill. I got my head out, and got some air. I tried to straighten my legs and crawl out, but my legs were numb.

It was still dark. I could not see what my captors looked like. I tried to massage my legs to get some life in them, and get my shoes on. I could hear now there were at least four of them. They were standing around me, and continuously chattering. I had never heard of Sasquatch before the Indian told me about them, but I knew I was right among them.

But how to get away from them, that was another question. I got to see the outline of them now, as it began to get lighter, though the sky was cloudy, and it looked like rain, in fact there was a slight sprinkle.

I now had circulation in my legs, but my left foot was very sore on top where it had been resting on my hobnail boots. I got my boots out from the sleeping bag, and pulled them on. I tried to stand up. I was wobbly on my feet but had a good hold of my rifle.

I asked: "What you fellows want with me."

Only some more chatter.

It was getting lighter now, and I could see them quite clearly. I could make out forms of four people. Two big and two little ones. They were all covered with hair and no clothes on at all.

I could now make out mountains all around me. I looked at my watch. It was 4:25 a.m. It was getting lighter now and I could see the people clearly.

"The Old Man"—Over 8 feet tall, 800 pounds.

They looked like a family, old man, old lady and two young ones, a boy and a girl. The boy and girl seemed to be scared of me. The old lady did not seem too pleased about what the old man dragged home. But the old man was waving his arms and telling them all he had in mind. They all left me then.

I had my compass and my prospecting glass on strings around my neck. The compass in my left hand shirt pocket and my glass in my right hand pocket. I tried to reason our location, and where I was. I could see now that I was in a small valley or basin about eight or ten acres, surrounded by high sides. There was a V-shaped opening about eight feet wide at the mountains, on the southeast bottom and about 20 feet high at highest point—that must be the way I came in. But how will I get out? The old man was now sitting near this opening.

I moved my belongings up close to the west wall. There were two small cypress trees there, and this would do for a shelter for the time being, until I find out what these people want with me, and how to get away from here. I emptied out my packsack to see what I had in the line of food. All my canned meat and vegetables were intact and I had one can of coffee. Also three small cans of milk—two packages of hard tack and my butter sealer half full of butter. But my prunes and macaroni were missing. Also my full box of shells for my rifle. I only had six shells beside what I had in the magazine of my rifle. I had my sheath knife but my prospecting pick was missing and my can of matches. I only had my safety box full and that held only about a dozen matches. That did not worry me—I can always start a fire with my prospecting glass when the sun is shining, if I got dry wood. I wanted hot coffee but I had no wood, also nothing around here that looked like wood. I had a good look over the valley from where I was—but the boy and the girl were always watching me from behind some juniper bush. I decided there must be some water around here. The ground was leaning towards the opening in the wall. There must be water at the upper end of this valley, there is green grass and moss along the bottom.

All my utensils were left behind. I opened my coffee tin and emptied the coffee in a dishtowel and tied it with the metal strip from the can. I took my rifle and the can and went hunting for water. Right at the head under a cliff there was a lovely spring that

"The Boy"—7 feet tall, 400 pounds.

"The Girl"—6 feet tall, 250 to 300 pounds.

disappeared underground. I got a drink and a full can of water. When I got back the young boy was looking over my belongings, but did not touch anything. On my way back, I noticed where these people were sleeping. On the east side wall of this valley was a shelf in the mountain side, with overhanging rock, looking something like a big undercut in a big tree about 10 feet deep and 30 feet wide. The floor was covered with lots of dry moss, and they had some kind of blankets woven of narrow strips of cedar bark, packed with dry moss. They looked very practical and warm—with no need of washing.

The first day not much happened. I had to eat my food cold. The young fellow was coming nearer me, and seemed curious about me. My one snuff box was empty, so I rolled it towards him. When he saw it coming, he sprang up quick as a cat, and grabbed it. He went over to his sister and showed her. They found out how to open and close it—they spent a long time playing with it—then he trotted over to the old man and showed him. They had a long chatter.

Next morning, I made up my mind to leave this place—if I had to shoot may way out. I could not stay much longer. I had only enough grub to last me till I got back to Toba Inlet. I did not know the direction but I would go down hill and I would come out near civilization some place. I rolled up my sleeping bag, put that inside my pack sack—packed the few cans I had—swung the sack on my back, injected a shell in the barrel of my rifle and started for the opening in the wall. The old man got up, held up his hands as though he would push me back.

I pointed to the opening, I wanted to go out. But he stood there pushing towards me—and said something that sounded like Soka, Soka. Again I pointed outside. He only kept pushing with his hands saying Soka, Soka. I backed up to about 60 feet. I did not want to be too close, I thought if I had to shoot may way out. A 30-30 might not have much effect on this fellow, it might make him mad. I only had six shells so I decided to wait. There must be a better way than killing him in order to get out from here. I went back to my campsite to figure out some other way to get out.

If I could make friends with the young fellow or the girl, they might help me. If I only could talk to them. Then I thought of a fellow who saved himself from a mad bull by blinding him with snuff in his eyes. But how will I get near enough to this fellow to put the snuff in his eyes? So I decided next time I give the young fellow my snuff box to leave a few grains of snuff in it. He might give the old man a taste of it.

But the question is, in what direction will I go, if I should get out. I must have been near 25 miles northeast of Toba Inlet when I was kidnapped. This fellow must have traveled at least 25 miles in the three hours he carried me. If he went west we would be near salt water—same thing if he went south—therefore he must have gone northeast. If I then keep going south and over two mountains, I must hit salt water some place between Lund and Vancouver.

The following day I did not see the old lady till about 4:00 p.m. She came home with her arms full of grass and twigs of all kinds from spruce and hemlock as well as some kind of nuts that grow in the ground. I have seen lots of them on Vancouver Island. The young fellow went up the mountain to the east every day, he could climb better than a mountain goat. He picked some kind of grass with long sweet roots. He gave me some one day—they tasted very sweet. I gave him another snuff box with about a teaspoon of snuff in it. He tasted it, then went to the old man—he licked it with his tongue. They had a long chat. I made a dipper from a milk can. I made many dippers—you can use them as pots too—you cut two slits near the top of any can—then cut a limb from any small tree—cut down back of the limb—down the stem of the tree—then taper the part you cut from the stem. Then cut a hole in the tapered part, slide the tapered part in the slit you made in the can, and you have a good handle on your can. I threw one over to the young fellow, who was playing near my camp. He picked it up and looked at it, then he went to the old man and showed it to him. They had a long chatter. Then he came to me, pointed at the dipper then at his sister. I could see that he wanted one for her too. I had other peas and carrots, so I made one for his sister. He was standing only eight feet away from me. When I had made the dipper, I dipped it in water and drank from it, he was very pleased, almost smiled at me. Then I took a chew of snuff, smacked my lips, said that's good.

The young fellow pointed to the old man, said something that sounded like "Ook." I got the idea that the old man liked snuff,

118

and the young fellow wanted a box for the old man. I shook my head. I motioned with my hands for the old man to come to me. I do not think the young fellow understood what I meant. He went to his sister and gave her the dipper I made for her. They did not come near me again that day. I had now been there six days, but I was sure I was making progress. If only I could get the old man to come over to me, get him to eat a full box of snuff that would kill him for sure and I wouldn't be guilty of murder.

The old lady was a meek old thing. The young fellow was by this time quite friendly. The girl would not hurt anybody. Her chest was flat like a boy—no development like young ladies. I am sure if I could get the old man out of the way, I could easily have brought this girl out with me to civilization. But what good would that have been? I would have to keep her in a cage for public display. I don't think we have any right to force our way of life on other people, and I don't think they would like it. (The noise and racket in a modern city they would not like any more than I do.)

The young fellow might have been between 11 and 18 years old about seven feet tall and might weigh about 300 lbs. His chest would be 50-55 inches, his waist about 36-38 inches. He had wide jaws, narrow forehead, that slanted upward round at the back about four or five inches higher than the forehead. The hair on their heads was about six inches long. The hair on the rest of their body was short and thick in places. The women's hair was a bit longer on their heads and the hair on the forehead had an upward turn like some women have—they call it bangs, among women's hair-do's. Nowadays the old lady could have been anything between 40-70 years old. She was over seven feet tall. She would be about 500-600 pounds.

She had very wide hips, and a goose-like walk. She was not built for beauty or speed. Some of those loveable brassieres and uplifts would have been a great improvement on her looks and her figure. The man's eyeteeth were longer than the rest of the teeth, but not long enough to be called tusks. The old man must have been near eight feet tall. Big barrel chest and big hump on his back—powerful shoulders, his biceps and upper arm were enormous and tapered down to his elbows. His forearms were longer than human's, but well proportioned. His hands were wide, the palm was long and

"The Old Lady"—7 feet tall, 600 pounds.

broad, and hollow like a scoop. His fingers were short in proportion to the rest of his hand. His fingernails were flat like chisels. The only place they had no hair was inside their hands and the soles of their feet and upper part of the nose and eyelids. I never did see their ears, they were covered with hair hanging over them.

If the old man were to wear a collar it would have to be at least 30 inches. I have no idea what size shoes they would need. I was watching the young fellow's foot one day when he was sitting down. The soles of his feet seemed to be padded like a dog's foot and the big toe was longer than the rest and very strong. In mountain climbing all he needed was footing for his big toe. They were very agile. To sit down they turned their knees out and came straight down. To rise they came straight up without help of their hands and arms. I don't think this valley was their permanent home. I think they move from place to place, as food is available at different localities. They might eat meat but I never saw them eat meat, or do any cooking.

I think this was probably a stopover place and the plants with sweet roots on the mountain side might have been in season this time of the year. They seemed to be most interested in them. The roots have a very sweet and satisfying taste. They always seem to do everything for a reason, wasted no time on anything they did not need. When they were not looking for food, the old man and the old lady were resting, but the boy and girl were always climbing something or jumping. The boy's favorite stunt was to take hold of his feet with his hands and balance on his rump, then bounce forward. The idea seems to be to see how far he could go without his feet or hands touching the ground.

Sometimes he made 20 feet.

But what did they want with me? They must understand I cannot stay here indefinitely. I will soon be out of grub and so far I have seen no deer or other game. I will soon have to make a break for freedom. Not that I was mistreated in any way. One consolation was that the old man was coming closer each day, and was very interested in my snuff. Watching me when I take a pinch of snuff. He seems to think it useless to only put it inside my lips. One morning after I had my breakfast, both the old man and the boy came and sat down only ten feet away from me. This morning I made coffee. I had saved up all dry branches I found and I had some dry moss and I used all the labels from cans to start a fire.

I got my coffee pot boiling and it was strong coffee too, and the aroma from boiling coffee was what brought them over. I was sitting eating hardtack with plenty of butter on, and sipping coffee. And it sure tasted good. I was smacking my lips pretending it was better than it really was. I set the can down that was about half full. I intended to warm it up later. I pulled out a full box of snuff, took a big chew. Before I had time to close the box the old man reached for it. I was afraid he would waste it, as I had only two more boxes. So I held on to the box intending him to take a pinch like I had just done. Instead he grabbed the box and emptied it in his mouth. Swallowed it in one gulp. Then he licked the box inside with his tongue.

After a few minutes his eyes began to roll over in his head, he was looking straight up. I could see he was sick. Then he grabbed my coffee can that was quite cold by this time, he emptied that in his mouth, grounds and all. That did no good. He stuck his head between his legs and rolled forwards a few times away from me. Then he began to squeal like a stuck pig. I grabbed my rifle. I said to myself, "This is it. If he comes for me I will shoot him plumb between his eyes." But he started for the spring; he wanted water. I packed my sleeping bag in my pack sack with the few cans I had left. The young fellow ran over to his mother. Then she began to squeal. I started for the opening in the wall—and I just made it. The old lady was right behind me. I fired one shot at the rock over her head.

I guess she had never seen a rifle fired before. She turned and ran inside the wall. I injected another shell in the barrel of my rifle and started downhill, looking back over my shoulder every so often to see if they were coming. I was in a canyon, and good travelling and I made fast time. Must have made three miles in some world record time. I came to a turn in the canyon and I had the sun on my left, that meant I was going south, and the canyon turned west. I decided to climb the ridge ahead of me. I knew I must have two mountain ridges between me and salt water and by climbing this ridge I would have a good view of this canyon, so I could see if the Sasquatch were coming after me. I had a light pack and was mak-

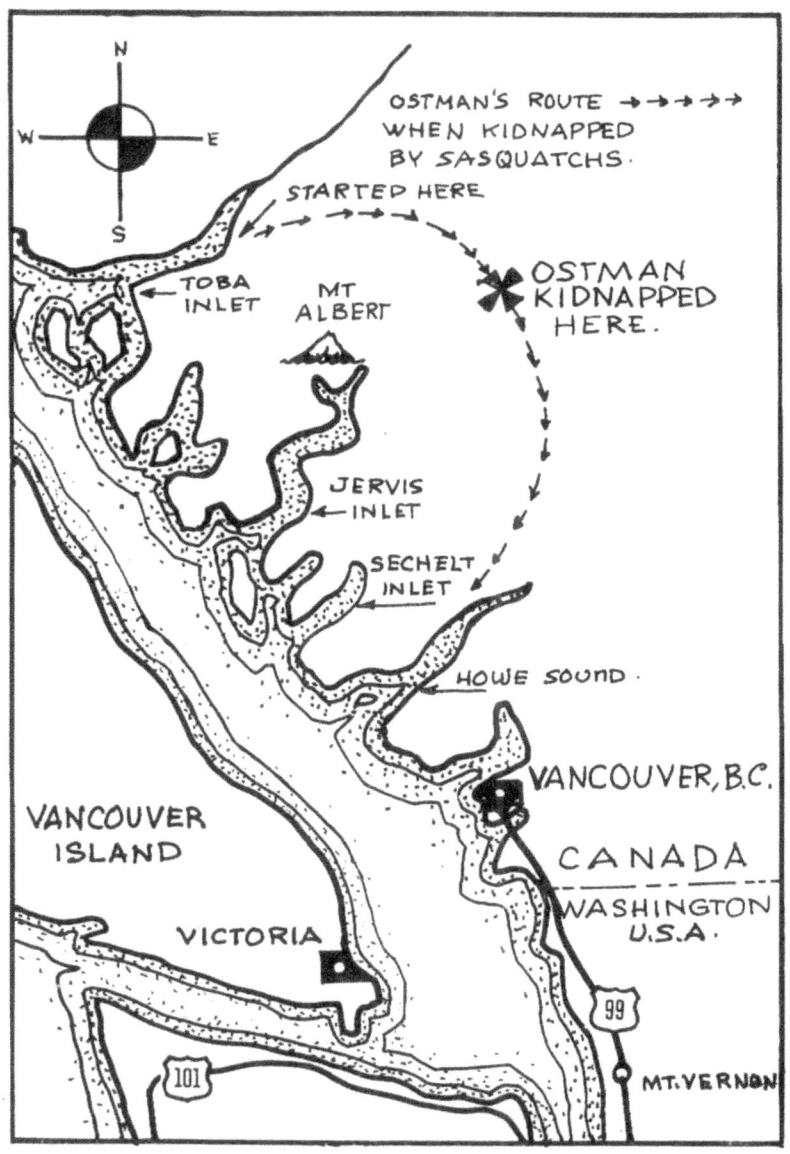

OSTMAN'S ROUTE →→→→→
WHEN KIDNAPPED
BY SASQUATCHS.

STARTED HERE

OSTMAN
KIDNAPPED
HERE.

TOBA
INLET

MT
ALBERT

JERVIS
INLET

SECHELT
INLET

HOWE SOUND.

VANCOUVER, B.C.

VANCOUVER
ISLAND

CANADA

WASHINGTON
U.S.A.

VICTORIA

99

101

MT. VERNON

ing good time up this hill. I stopped soon after to look back to where I came from, but nobody followed me. As I came over the ridge I could see Mt. Baker, then I knew I was going in the right direction.

After reading Mr. Ostman's story, I went to Canada to personally interview him. His encounter with the Sasquatch would be worthy of a book in itself. I came away with a firm belief that his story was true.

I also contacted Mr. John Green at Agassiz while in British Columbia. Mr. Green has collected a great deal of information on the Sasquatch, some of which he has printed in the Agassiz Advance newspaper which he owns and edits. Here are a few newspaper articles from that publication, plus some Indian stories:

BERRY PICKER MET HAIRY GIANT NEAR NELSON, B. C.

Many people are under the impression that stories about Sasquatches are to be found only in Indian legends. This is not so, there are many accounts in which no Indians are involved at all, and by no means all of these stories date from earlier days. The most recent, to our knowledge, was printed in the Nelson News, October 4, 1960. The full text follows:

Man or beast or both? Whatever it was that sent John Bringsli of Nelson fleeing in blind panic from the head of Lemmon Creek, hurling his huckleberry pail into the bush and racing for home in his early-model car, it had pulled a speedy disappearing act by the time he and a group of hunters returned to the scene.

Mr. Bringsli, woodsman, hunter and fisherman in Kootenay district for more than 35 years, swore on his reputation as an outdoorsman that it was "definitely not a bear."

In an interview, Mr. Bringsli related his experience with an "unknown creature" seen while on a huckleberry picking expedition alone near Six-Mile and unashamedly told of his frantic race over 100 yards of stunted bush and dwarfed underbrush to his car.

"I had just stopped my 1931 coupe on a deserted logging road a couple of weekends ago and walked about 100 yards into the bush. I was picking huckleberries.

"I had just started to pick berries and was moving slowly through the bush. I had only been there about 15 minutes.

"For no particular reason, I glanced up and that's when I saw this great beast. It was standing about 50 feet away on a slight rise in the ground, staring at me."

"The sight of this animal paralyzed me. It was seven to nine feet tall with long legs and short powerful arms with hair covering its body. The first thing I thought was . . . what a strange looking bear."

"It had very wide shoulders, and a flat face with ears flat against the side of its head. It looked more like a big hairy ape."

"It just stood there staring at me. Arms of the animal were bent slightly and most astounding was that it had hands . . . not claws.

"It was about 8 a.m., and I could see it very clearly," Mr. Bringsli said.

The most peculiar thing about it was the strange bluish-grey tinge of color of its long hair. It had no neck. Its ape-like head appeared to be fastened directly to its wide shoulders."

Mr. Bringsli stood with mouth agape, staring at the thing for about two minutes. Then it began to slowly walk, or rather shuffle, toward the paralyzed huckleberry hunter. It was then that Mr. Bringsli decided it was time for him to find another berry picking location.

He sprinted the 100 yards to the car and drove recklessly down the old logging road and home.

Mr. Bringsli returned to the scene the next day with a group of friends armed with high-powered rifles and cameras but the strange beast did not reappear. They did find one track nearby. It was from 16 to 17 inches long. There were no claw marks but rather a "sharp toe" print as described by Mr. Bringsli.

When asked if he would return to that area again, he retorted, "Of course, but this time I'll take along the old 30.06 just for good luck."

WHITE MAN CAN'T SEE SASQUATCH—
HE DOESN'T SMELL GOOD ENOUGH

The following letter was received by Village Clerk Paul Trout in 1957 when Harrison Hot Springs, over the protests of Commissioner Bob Gill, was planning a search for the Sasquatch as a Centennial project.

Katz, B.C., April 23

Dear Mr. Trout:

Maybe you white man like Mr. Gill don't believe in Sasquatch but there are lots of things you don't know. Fifteen years ago my old daddy was hurt bad by Sasquatch man he met a mile from Katz. Some white man say my daddy must be drunk when he get his arm broke but no Indians laff. Only whites when they don't know nothing laff like fools.

One thing my daddy was good Catholic and he very little drink likker. White man's poison he say.

What happens he say was daddy was with momma picking berries when he went away from others for rest. He say he only look at trees and sky, then big man over six foot comes running from rocks at him, hit old daddy to ground, hit him on head and side and arm, hit him hard and make grunts. Daddy yell then others come and Sasquatch run away fast. They see Sasquatch running and daddy blood on his head.

Grandma Charlie set bones in arm she say that little runty Sasquatch maybe other Sasquatch treat him mean so he treat little Indian like old Daddy mean.

Grandma say Sasquatch big nice man is catch little Indian woman for make love to all they want. Old daddy scared of woods after, never go anywhere, just stay home. He cross river springtime in old boat but never go in trees, maybe mean old Sasquatch hiding there, hair all over and deerskin. I think maybe daddy drink beer and forget what Sasquatch do, but he still good guy, not work but good. He have crooked arm till he die two years in 1955.

You white man know lot smart things, smart guys, big car, big house, but you not smart in everything. Maybe Sasquatch hate white man smell and not show. Grandma Charlie say white man

126

smell like old dead man and scare Sasquatch. You maybe see Sasquatch. You go in woods. Catch one if you take Indian. Make Indian go in front with salmon or deer. Leave on tree branch, then catch. Sasquatch eat lots, always want eat. I hope you catch, then you know Indian tell straight. Good luck.

Mary Joe

How powerful are these giants? A group of Indians came upon a small canyon in British Columbia, and were petrified when they saw a huge hairy giant and a large brown bear in an ear-shattering battle. It was a long, hard fight, but the giant finally strangled the big bear to death!

In 1963, I read a very interesting story that was published in Sports Afield Magazine, entitled "Long Hunter—Alaskan Style" by Russell Annabel. The story is about Tex Cobb, a mountain man who spent years trapping in Canada and Alaska. The last half of the article reported:

The Denna Indian people liked him, Tex Cobb. No sentiment was wasted on either side, but he and the tribesmen had a live-and-let-live understanding that was rare in those days. He stayed off their trap lines, and they stayed off his. If an Indian had a salmon net in an eddy, Tex found another eddy, and vice versa. Due to the fact that the Indians trusted him, we became involved one autumn with what today would be called, I suppose, an abominable snowman. I have since heard and read a great deal about the abominable snowman. I have seen the photographs of those tracks in the snow on a Tibetan mountain, and to me they are simply the tracks of a man with gunnysacking or some other cloth wrapped around his feet as protection from the cold, climbing slewfooted because the slope was steep and he had no crampons. But when I was a youngster roaming the North with Tex, we had never heard of the abominable snowman. We had, however, heard much about Gilyuk, the shaggy cannibal giant sometimes called The-Big-Man-With-The-Little-Hat.

Our adventure with Gilyuk occurred while we were camped in a pretty spruce park on Yellowjacket Creek, south of Tyone Lake. We had spent the entire summer on this mountain-girt Nelchina Plateau, wandering about in aimless nomad fashion. Tex said we were prospecting and looking for fur sign. Maybe we were. He

Brown bear, king of the North, meets his match.

Someone's going to get hurt.

always had to have an excuse for enjoying the country, a commercial excuse if he could think of one. Anyway, it was now late September, the beautiful time, no mosquitoes, the land ablaze with color, the fish and the meat animals summer-fat, the caribou horde gathering, and we were footloose and free as perhaps men can never be again. This morning Tex was making coffee, and I was down at the creek cleaning a mess of grayling for breakfast, when six Indians filed in through the timber. They stood a moment solemnly regarding our four horses. To them a horse was a rarity, a mysterious animal. They called them McKinley moose, because McKinley was the only president they had ever heard of, and the horses were as big as moose. I followed them to the camp.

"Have you eaten?" Tex asked them in Denna.

They said they had eaten. Chief Stickman was with them. I had seen him once before, at Eklutna Village. A squat, square-faced man, very dark, with long hair and quick-moving obsidian eyes, he was the Denna boss of this entire area, and his reputation was bad. But now he had trouble that he couldn't handle. He told us about it, and as he talked, he kept standing first on one leg, then the other, balancing himself with the moccasined sole of the free foot against the knee of the supporting leg. I don't know whether it was habit or a medicine trick to ward off evil spirits, or both, but it was disconcerting. He had come into this area two days ago, he said, with some of his people to kill and cache caribou for winter use. But they had discovered that Gilyuk, the shaggy giant, was hanging around. They had found his sign yesterday. And of course everybody knew that Gilyuk wasn't interested in caribou. Gilyuk ate men.

"What kind of sign?" Tex asked.

"We will take you to see it," Stickman said. "It is not far."

After breakfast we followed the Indians upstream a couple of miles to a burned flat on which a nurse crop of aspen and birch had grown. In the center of the flat stood a ruined birch sapling. It had been about four inches through and maybe ten feet tall. Something had twisted the sapling as a man would twist a match stick. The wood had separated into individual fibers, the bark hung in tatters. Stickman and his hunters stood back, while Tex and I looked the site over. Moose often ride a sapling down to get at the tender upper twigs. So do caribou. But no moose or caribou had done this. This had been done by something with hands. It had happened yesterday, because the leaves of the sapling had not yet completely wilted. It wasn't the work of lightning—no burns. A freak whirlwind hadn't done it, because trees and brush a few yards distant were undamaged. The hard ground showed no tracks. We found no snagged hair in the brush. Absolutely nothing except the incredibly twisted birch sapling. It was without question the eeriest sight I ever beheld in the wilds.

Stickman said, "It is Gilyuk's mark. We have seen it before."

I wish to make clear that to the Denna people Gilyuk was no legendary creature their grandfathers had told them about. He was a reality, and they spoke of him as they spoke of the bears and wolves. They saw his sign, and they saw him. He was a shaggy giant who wore a little hat and ate men. "We want to ask you to camp with us until we have killed our caribou," Stickman said. "Gilyuk doesn't molest white men. Perhaps he will not molest us if you are in the camp." Stickman had already told us that he was bivouacked on the shore of a pothole lake two hours to the eastward.

Tex said all right, we would move to his camp in the morning. As he spoke, he was still looking at the twisted sapling, his green eyes narrowed in thought. I couldn't take my gaze off it either.

Stickman said, "Thanks, Kosaki," a strange word of respect, held over from the old Russian Cossack, and we parted company with the Indians.

Next morning I brought the horses in at daybreak. We ate, broke camp and were putting on the packs, when here came the Indians, all of them—all, that is, except Stickman. An old man told us that they were returning to their town on Tyone Lake. Stickman was dead, he said. Gilyuk had taken him. The chief had got up in the night and gone down to the lake, perhaps for water, but nobody knew. A squaw with a birch-bark torch found his red flannel underwear on the gravel beach. It had been torn off him. There may have been tracks, but the entire hunting party had swarmed over the beach, and by daylight no tracker on earth could have made sense of the jumble.

Well, until the day of his own death last July, while on a senti-

131

mental journey to a fateful spot in Cook Inlet, Tex was convinced that the cannibal giant Gilyuk killed Stickman. When asked if he believed in the existence of abominable snowmen, Tex would reply that he didn't think there were any around in Alaska nowadays, but that they had existed, at least one of them, a couple of decades back. This is good enough for me. I go along with it.

The end of the line for a terrified trapper.
(Teddy Roosevelt told this story)

Here is a frightening story, from Idaho, told by none other than Theodore Roosevelt. The story was in a book he published in 1893 entitled, "Wilderness Hunter." Teddy spent a lot of time in the wilderness and he was a hard man to fool with a wild tale. This story seemed to have impressed him greatly mainly because of the still noticeable tremor of the voice of the old mountainman as he related the story to Teddy—even though it was a half a lifetime after it had happened. Roosevelt's strange story goes as follows:

It was told (to me) by a grizzled, weather-beaten old mountain hunter, named Bauman, who was born and had passed all his life on the frontier. He must have believed what he said, for he could hardly repress a shudder at certain points of the tales.

When the event occurred Bauman was still a young man, and was trapping with a partner among the mountains dividing the forks of the Salmon from the head of Wisdom River. Not having had much luck, he and his partner determined to go up into a particularly wild and lonely pass through which ran a small stream said to contain many beaver. The pass had an evil reputation because the year before a solitary hunter who had wandered into it was there slain, seemingly by a wild beast, the halfeaten remains being afterwards found by some mining prospectors who had passed his camp only the night before.

The memory of this event, however, weighed very lightly with the two trappers, who were as adventurous and hardy as others of their kind. . . . .They then struck out on foot through the vast, gloomy forest, and in about 4 hours reached a little open glade where they concluded to camp, as signs of game were plenty.

There was still an hour or two of daylight left, and after building a brush lean-to and throwing down and opening their packs, they started up stream. . . .

At dusk they again reached camp. . . .

They were surprised to find that during their absence something, apparently a bear, had visited camp, and had rummaged about among their things, scattering the contents of their packs, and in sheer wantonness destroying their lean-to. The footprints of the beast were quite plain, but at first they paid no particular heed to

135

them, busying themselves with rebuilding the lean-to, laying out their beds and stores, and lighting the fire.

While Bauman was making ready supper, it being already dark, his companion began to examine the tracks more closely, and soon took a brand from the fire to follow them up, where the intruder had walked along a game trail after leaving the camp. . . . Coming back to the fire, he stood by it a minute or two, peering out into the darkness, and suddenly remarked: "Bauman, that bear has been walking on two legs." Bauman laughed at this, but his partner insisted that he was right, and upon again examining the tracks with a torch, they certainly did seem to be made by but two paws, or feet. However, it was too dark to make sure. After discussing whether the footprints could possibly be those of a human being, and coming to the conclusion that they could not be, the two men rolled up in their blankets, and went to sleep under the lean-to.

At midnight Bauman was awakened by some noise, and sat up in his blankets. As he did so his nostrils were struck by a strong, wild-beast odor, and he caught the loom of a great body in the darkness at the mouth of the lean-to. Grasping his rifle, he fired at the vague, threatening shadow, but must have missed, for immediately afterwards he heard the smashing of the underwood as the thing, whatever it was, rushed off into the impenetrable blackness of the forest and the night.

After this the two men slept but little, sitting up by the rekindled fire, but they heard nothing more. In the morning they started out to look at the few traps they had set the previous evening and put out new ones. By an unspoken agreement they kept together all day, and returned to camp towards evening.

On nearing it they saw, hardly to their astonishment, that the lean-to had been again torn down. The visitor of the preceding day had returned, and in wanton malice had tossed about their camp kit and bedding, and destroyed the shanty. The ground was marked up by its tracks, and on leaving the camp it had gone along the soft earth by the brook, where the footprints were as plain as if on snow, and, after a careful scrutiny of the trail, it certainly did seem as if, whatever the thing was, it had walked off on but two legs.

The men, thoroughly uneasy, gathered a great heap of dead logs, and kept up a roaring fire throughout the night, one or the other sitting on guard most of the time. About midnight the thing came down through the forest opposite, across the brook, and stayed there on the hill-side for nearly an hour. They could hear the branches crackle as it moved about, and several times it uttered a harsh, grating, long-drawn moan, a peculiarly sinister sound. Yet it did not venture near the fire.

In the morning the two trappers, after discussing the strange events of the last 36 hours, decided that they would shoulder their packs and leave the valley that afternoon. . . .

All the morning they kept together, picking up trap after trap, each one empty. On first leaving camp they had the disagreeable sensation of being followed. In the dense spruce thickets they occasionally heard a branch snap after they had passed; and now and then there were slight rustling noises among the small pines to one side of them.

At noon they were back within a couple of miles of camp. In the high, bright sunlight their fears seemed absurd to the two armed men, accustomed as they were, through long years of lonely wandering in the wilderness to face every kind of danger from man, brute, or element. There were still three beaver traps to collect from a little pond in a wide ravine near by. Bauman volunteered to gather these and bring them in, while his companion went ahead to camp and made ready the packs.

On reaching the pond Bauman found 3 beavers in the traps, one of which had been pulled loose and carried into a beaver house. He took several hours in securing and preparing the beaver, and when he started homewards he marked, with some uneasiness how low the sun was getting. . . .

At last he came to the edge of the little glade where the camp lay, and shouted as he approached it, but got no answer. The camp fire had gone out, though the thin blue smoke was still curling upwards. Near it lay the packs wrapped and arranged. At first Bauman could see nobody; nor did he receive an answer to his call. Stepping forward he again shouted, and as he did so his eye fell on the body of his friend, stretched beside the trunk of a great fallen spruce. Rushing towards it the horrified trapper found that the body was still warm, but that the neck was broken, while there were four great fang marks in the throat.

The footprints of the unknown beast-creature, printed deep in the soft soil, told the whole story.

The unfortunate man, having finished his packing, had sat down on the spruce log with his face to the fire, and his back to the dense woods, to wait for his campanion. . . . It had not eaten the body, but apparently had romped and gambolled round it in uncouth, ferocious glee, occasionally rolling over and over it; and had then fled back into the soundless depths of the woods.

Bauman, utterly unnerved, and believing that the creature with which he had to deal was something either half human or half devil, some great goblin-beast, abandoned everything but his rifle and struck off a speed down the pass, not halting until he reached the beaver meadows where the hobbled ponies were still grazing. Mounting, he rode onwards through the night, until far beyond the reach of pursuit.

There are many other states in the United States that have reported giant creatures that roam about their mountain wildernesses.

However, I do not have enough verified information to fully go into it at the present time. Anyway, that would be another book.

Chapter 5

BIGFOOT, GIANT HAIRY APE, SASQUATCH— PAST, PRESENT AND FUTURE

Do you still have some doubts about the actual existence of these huge pre-historic-type man creatures? If you do, perhaps this final chapter will answer some of your questions.

Once more, I must refer to Ivan Sanderson and his article, "Abominal Snowmen Are Here!"

ABOMINABLE SNOWMEN ARE HERE!

By IVAN T. SANDERSON

Sooner or later somebody always asks me: "You don't really think there is such a thing as an Abominable Snowman, do you?" My reply is always the same: "No. I believe there are hundreds if not thousands of unknown anthropoids, of at least half a dozen kinds, running all over five continents." And I usually add for good measure: "But they're not men, none of them lives in snow, and we have no right to call them abominable."

That has ended a great many conversations, and not a few friendships, but with all the evidence which has became available over the years, I feel it's almost in a league with asking me if I really believe that the earth is round.

To begin with, let us dispose of the ridiculous title "Abominable Snowman." It is a complete misnomer and extremely misleading. Worse, it is usually prefixed with the article "the," just as if there was but one lone, mateless, childless and parentless monster that has been pounding about the eastern Himalaya and south Tibetan upper snowfields for 50 years—a forlorn abomination, left over from the past or, perhaps, just spontaneously created out of the mists.

As to the adjective "abominable," I don't think we can call any living creature by that name. The things are probably quite decent; just scared, and demanding only that they may lead their lives in peace. Whether they may be called men is also debatable. In my opinion, some are and some aren't. I am firmly convinced that they range from extremely primitive humans, without true speech, tools or knowledge of fire-making, and still in varying degrees hairy, to one or two still undiscovered large apes in Africa. In between, some appear definitely to be Neanderthal submen such as inhabited Europe in the ice age but which have lingered on in eastern Asia, while others are even farther down the Hominid (Man) branch of the family tree, being what used to be called "Ape-men."

Then quite different from all these, there is the creature the Nepalese call the Meh-Teh, the original "Snowman," and which is to science truly abominable. By the footprints it leaves and all the descriptions of it and its behavior by eyewitnesses, it is the most bestial of all. What is more, the tracks it leaves are not Hominid, Pangid (Ape-like), or even really anything in between. They are quite unlike anything we know, dead or alive.

But it is the word *snow* that bugs the whole business. Since words are intended to convey meaning, nobody can be accused of stupidity for supposing that this title is intended to indicate either a man made of snow, or a man that lives in or on snow. Since nobody seems dense enough to believe the former, one can only assume the latter.

But this, too, is ridiculous. Many of these tracks have been found on permanent mountain snowfields, and there is nothing at all under these snowfields which could sustain any living creature. While they cross these snowfields when going from one place to another, thus leaving the tracks which have been seen by Sir Edmund Hillary, among others, they actually live in the forests which, admittedly, often come right up to the snowline.

Having thus, I hope, disposed of the business which has done more than anything else to muddle the whole issue, I will now proceed to answer your second question: "But how on earth could there be such creatures running about all over the lot?"

141

This is a very good question because it can be easily disposed of. First, a very large part of the land surface of our earth is uninhabited. A considerable part of this is still unmapped, unused, and has not even been explored. About a seventh of it is said to be covered with permanently frozen soil, and over most of this, which is in the Arctic and sub-Arctic, there sprawls an endless forest of tightly packed spruce trees known as the taiga. This runs right around the top of the world from northern Russia, through Siberia, to the Bering Straits, and then picks up again on the lowlands of the Canadian Northwest Territories and continues unbroken right across our continent to Labrador. It is virtually uninhabited, and only in the two last decades have roads been driven into it.

Of the remainder of the land surface, a third is either uninhabitable hot desert or its surrounding scrublands. Little of the latter is permanently settled, and the major part is totally unused and seldom crossed. Of the remainder, nearly half is covered with forests. Although some of these forests are dotted with human settlements, they are mostly what we call wildernesses, and most of them are unmapped. People get lost in Maine every year, and there is a 15,000-square-mile block in northern California that is only just being surveyed. There are areas of over 1,000 square miles in the Mississippi Valley bottomlands that are crossed by only one third-class road and can show but half a dozen settlements.

There are great tracts even in old Europe that are complete wildernesses, but even more fantastic are the uninhabited blocks in subtropical and tropical countries like southern China proper and India, which we think of as positively bulging with population. And Communist officials empowered to look after "minorities" in China reported only five years ago that completely wild, hairy people without speech, clothes, tools or knowledge of fire had been captured in the border province of Yunnan and taken to the capital city of Kunming.

There is another reason why I am so certain that "Abominable Snowmen" can be existing in many areas of the world. This is due to the fact that many huge creatures have been discovered, and even in regions where the local people had no idea that they existed. In 1960, for example, the regular "Mountie" air-patrol spotted in the Canadian Northwest Territories, not 100 miles from the

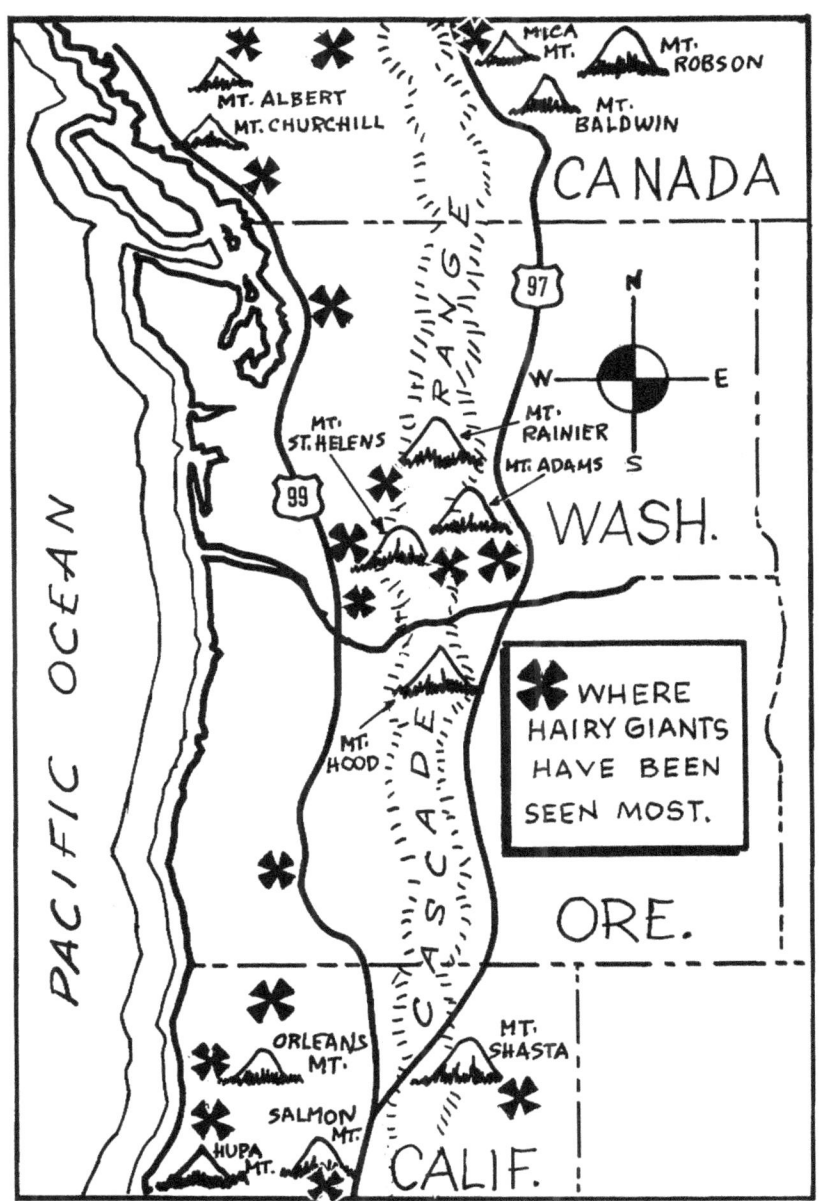

MICA MT.
MT. ROBSON
MT. ALBERT
MT. CHURCHILL
MT. BALDWIN
CANADA
97
N
W   E
S
MT. ST. HELENS
MT. RAINIER
MT. ADAMS
99
WASH.
WHERE HAIRY GIANTS HAVE BEEN SEEN MOST.
MT. HOOD
CASCADE RANGE
ORE.
PACIFIC OCEAN
ORLEANS MT.
MT. SHASTA
SALMON MT.
HUPA MT.
CALIF.

new road being pushed up to the Arctic Ocean from Alberta, and within 50 miles of a Mission Station established a century ago, large herds of what is either the second or third largest form of the Ox Tribe.These were groups of pure-blood Woodland Bison (Bison Athabasca), an enormous ice-age species not known to exist in a pure strain anywhere.

This was bad enough, but at least it was in the seemingly endless Taiga forest. In 1938, however, another creature—also either the second or third largest member of the Ox Tribe—turned up in the thickly populated Indochinese Peninsula. This creature is quite fabulous, the males having wide-spreading horns like the extinct Aurochs of Europe, father of all our western domestic cattle, but with huge tassels sprouting upwards from about a foot below their tips. What is even more significant, the discovery of this animal was at first positively denied in scientific circles although the man responsible took a complete skin and skull to Paris.

I could go on and on: the Coelacanth fishes, thought to have been extinct for 60 million years, turning up on the breakfast tables of the Comorro Islanders; the second largest land mammal, named Cotton's Ceratothere or White Rhinoceros found only in 1910; the forest giraffe or Okapi of the Congo in the same year, and so on. But what is the use? May not these two sets of facts— the general unexplored nature of our earth, and the discovery right up till now of herds of huge beasts right at our back doors—suffice to affirm my contention that many undetected creatures can still be existing almost in our midst?

At this point, I believe you will be saying to yourself: "Yes, this is all very well, but those are real animals. These snowmen are nothing but stories, however important and reliable the people who have told these stories may be. Is there any concrete physical evidence of their existence?"

The answer is a definite "Yes."

I think we will have to admit that foot tracks are fairly concrete, so let's begin by taking another look at those of ABSMs (as I will refer to them from now on) and at the circumstances in which thy were found.

Footprints can appear in all manner of soft and resilient surfaces; both dry, like sand and gravel; or wet, like mud and snow.

144

Despite all the folderol about those found in snow, far more have been found in mud and sand, and of course exclusively so in all lowland areas in subtropical and tropical lands. The story of their discovery is seldom dramatic, but when it is, it is exceedingly so.

The best known is undoubtedly that of the famed mountaineer, Eric Shipton, in the Everest Area in 1952. The next most familiar is the California affair which I reported in *True* in December 1959. In both cases it was not, however, so much the incident itself that made such an impression, but the definite and more or less unassailable proof that was obtained at the time in the form of photographs and plaster casts, plus the fact that in both areas the tracks were seen by several people at the same time.

Further, these people were educated men with reputations of the highest order. Yet, the world at large was not ready at either time for such an event, nor was the public in any way prepared to accept it.

Eric Shipton was exploring a range of mountains near the Everest Block named the Gauri Sankar on the South Tibetan Rim. He was accompanied by one Michael Ward and the Sherpa, Sen Tensing. On the afternoon of the 8th of November they stumbled upon a fresh track made by a Meh-Teh. This was in powdery snow on the southwestern slope of the Menlung-tse. The individual imprints were absolutely clear-cut. Their maker walked on two feet. The track was followed up for over a mile to an ice moraine, into which the men could not follow. The Meh-Teh had jumped some crevasses and had dug its toes in to do so just as any human would.

The tracks and prints were photographed, and the form of these prints and the stride of the track corresponded with similar discoveries of dozens of others, both previously and since. The photographs, and molds based on them, were exhibited in London alongside those made by bears and a large monkey. I may add that the keynote of this exhibit was "Now you can see for yourself that these so-called abominable snowman tracks are only those of a bear—or a monkey!" If you will compare the tracks pictured with this article with those of a bear or a monkey you will see how ridiculous this is. How even a stay-at-home scientist in a museum could be so stupid, I fail to understand.

145

Comparative Foot Sizes

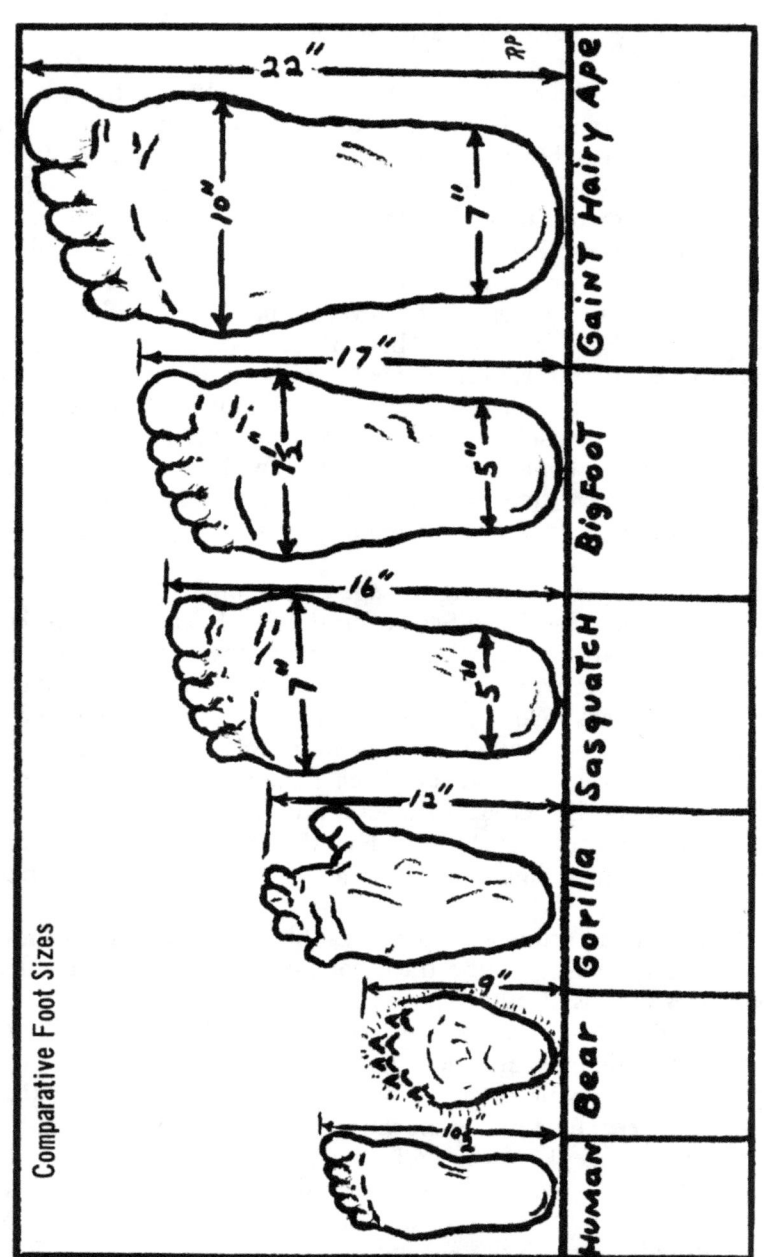

Human	Bear	Gorilla	Sasquatch	Bigfoot	Gaint Hairy Ape

The California affair was altogether different. There, enormous footprints turned up night after night all over a new road being bulldozed into a wilderness area not 100 miles from the town of Eureka. They were inspected by several dozen hard-boiled and highly practical-minded bulldozer operators, loggers, road-engineers and even by press photographers. They were up to 22 inches long, appeared night after night out of the impenetrable forests, went up and down impossible slopes, meandered around the machinery left parked at night, and then wandered off back into the wild with 60 inch strides. They caused a great stir which prompted some enquiry. This brought to light the fact that such things had been reported off and on for a century all over the area and as far away as Idaho, Oregon and Washington. Further, they linked up with similar sightings in British Columbia.

Tracks, which play such an important part in the whole business of ABSMery, have been found all over the world. Several of the casts made from these tracks are so clear and perfect that the musculature of the bottom of the feet that made them has been worked out in detail. In some types this proves to be very human in form, as with the so-called Almas of northeastern asia. In others, it is absolutely not human, as in the Meh-Teh. In the Oh-Mahs of California, with a second pad under their big-toes and their apparent webbing of all the toes up to the last joints, we have something almost—but not quite—human.

This is really quite an impressive showing and when we came to properly appreciate the fact that tracks have been reported by Mongolians, Chinese, Nepalis, Tibetans, Russians, Persians, Africans, Malays, Hollanders, Belgians, and members of most other European nationalities all over the world, and by Canadians and other North, Central and South Americans—year in and year out for over a century, it becomes very hard to see how anybody can really doubt the existence of ABSMs.

Also, you will have to admit that such theories as the hoax, the mis-identification, the tall-tale or the pure lie become so utterly ridiculous that they are not worth even discussing. The real trouble about them, however, is that they are just about the only points of view you ever read on the matter. And one and all, they are nothing more than attempts to disprove the whole business by trying

147

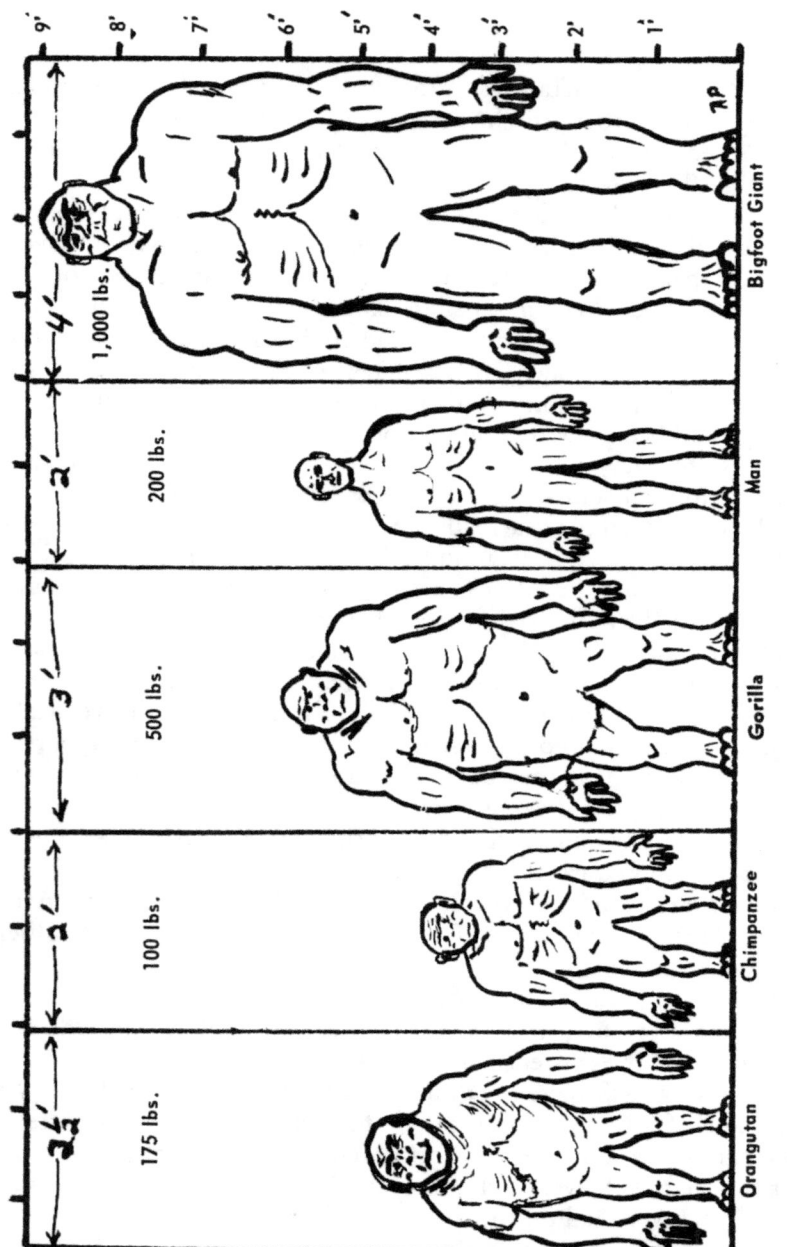

148

to debunk one small aspect of it. No better example of this can be given than the circus put on by Sir Edmund Hillary early this year.

Hillary stated in an article published before he went to the Himalayas that his secondary objective was to get what he called a yeti (an over-all local misnomer for any ABSM). He is a mountaineer, not a zoologist or anthropologist, so he went clean through the country in which the ABSMs live and up on to the sterile, foodless, mountain snowfields. Also, he had a party of no less than 600 along with him. Failing, as a result, to get within miles of any ABSM, he was faced with two choices: either admit failure, or somehow disprove the whole idea.

He chose the latter course; had a "scalp" made from a skin taken from a rare local animal named a Serow; borrowed one of the old caps made to look like an ABSM scalp (and which was admitted to be a fake by the villagers he got it from); invited a most excellent Nepalese gentleman named Kunyo Chumbi to come along and flew off around the world, displaying the cap on television and handing out hairs and bits of skin to scientists. With these bits went a challenge to identify the hairs and dried blood. It took a scientist in Paris just one day to identify the hairs (as being from a Serow) but, strangely, microphotos of them did not match those made of hairs pulled from other scalps in Nepal by other scientists! On the basis of this confusing and meaningless test Sir Edmund presumed to claim that no ABSMs existed.

Then Hillary was asked: if the debunking of this scalp disproved the existence of all ABSMs everywhere, how about the tracks which he himself had several times reported? To this he produced the amazing reply that they were all made by a string of foxes following a leader and all landing precisely with all their feet in exactly the same hole, and then all these holes being enlarged by melting precisely to the same size and shape. (We have been unable to trace any reference to any species of fox ever being collected in upper montane regions.)

To the two questions, how then did these tracks invariably show not only clear toe marks of a very special arrangement, but also distinctive musculature impressions, and how could such tracks be made in mud which does not melt, he gave no answer!

Equally fatuous was the suggestion made by a gentleman named

149

Michael Peissel who wrote that the tracks in the Himalayas were made by men wearing a kind of mukluks, which had worn out in front so that they left toe impressions. He further said that such tracks are deliberately pointed out by the Nepalese as a tourist attraction! Should this be so, even in that area, all said men must have had both feet constructed in one of the rarest known ways— an abnormality in which the second toes are longer than the first, and are also bigger, and separated from the others; while they must all have been positively enormous people with feet almost as wide as long, and all have been twice the weight of a normal large man.

Apart from tracks, the physical evidence for the existence of ABSMs consists of a few alleged scalps (and they are definitely not all made from from the skins of goat-like animals), a few whole skins reported by Mongolian scientists; some mummified hands; several collections of fresh droppings; a lot of hairs; some analyses of old blood; and the identification of some odd internal and external parasites taken from said scalps and droppings.

Apart from this, everything is reportage—of weird calls made by, appalling smells from; animals found killed by; cairns on mountain tops being moved by; rocks being hurled by; beds being made by, and a few other minor categories.

Perhaps the most concrete evidence we have are two or three mummified hands. Two are preserved in a monastery in a small place in Nepel called Pangboche. There is a great mystery about one of these because it has only been photographed once, but then by one of the greatest students of the subject with the very highest standing—Prof. Teizo Ogawa, of the School of Anatomy of Tokyo University. It is the most perfect shot and shows some significant features. Professor Ogawa has not yet completed his examination of it, nor published his report, so that he has made no final pronouncement on its identification.

The other hand has now definitely been pronounced, and by none other than Prof. B. F. Porshnev, head of the Special Commission to Study ABSMs set up by the Soviet Academy of Sciences to be that of a Neanderthal subman, such as inhabited Europe and northern Asia during the last ice advance. Significantly, a fresh footprint from central Asia of a form of ABSM called a Guli-

Yavan almost exactly matches one left in an Italian cave some 50,000 years ago by a Neanderthaler. The cave got sealed by a stalactite curtain and, when broken into in 1952, these tracks were found, as fresh as if they had been made the day before, in the clay covering its floor.

The other most definite and concrete evidence we have is the scat or droppings. This constitutes a substance that cannot be manufactured or faked. And in several cases there was no other animal known that could deposit them. Also parasites found in these droppings have been found to be odd in several respects, notably that some are known only from animals, some from human beings, and others from nowhere previously. The same goes for certain mites taken off the scalps and other hairy bits of ABSMs examined.

This brings me to the question I know you have been hankering to ask: Then, why hasn't anybody seen one?"

This question often crops up in newspaper accounts and articles on the subject in a rather glib form, such as: "These creatures, never seen by a white man . . . (etc.)" This to me is an astonishing statement because there are literally dozens of reports of all the different kinds having been seen all over the world, and by all manner of people from the humblest peasants and most primitive tribesmen to military doctors in the Soviet Army, famous British mountaineers, and even roving American scientists. In fact, there are as many cases of "sightings" on record as there are of tracks.

The whole business, indeed, was kicked off in modern times by a very definite sighting. This was made by none other than the famous explorer and mountaineer, Col. C. K. Howard-Bury, when on the first real attempt to climb Mount Everest in 1921.

On November 21 of that year the party was on the way from a place named Kharta to the famous Lhapka-la Pass when somebody spotted a number of large dark objects moving about on a high snowfield well above them and at some distance. These were observed by the whole party and through binoculars, but they were too far distant to identify. When the mountaineers reached the area on the next afternoon they found large numbers of huge tracks which they described as being "three times as big as a normal footprint."

They were obviously left by some creature walking on its two hind legs, but Colonel Bury later said he thought that they had been made by "a large stray gray wolf!" The Sherpa porters disagreed, saying definitely that it had been a party of Meh-Tehs, and this name got garbled by an Indian telegraphist and came out as Metoh-Kangmi. This, an Englishman in India said, was Tibetan for "Abominable Snowman." (The expression happened to be Nepali, and the Englishman did not speak either that language or Tibetan, but let it pass.)

Actually, there had been others in that general area who had reported seeing the same or similar types of creatures. There is a gentleman by the name of Hugh Knight who is supposed to have met one face to face. It was shaggy and carried a crude bow and arrow. Then there was the famous botanist-explorer named Elwes who reported to the Zoological Society of London that he had seen one run over a ridge in 1916.

After Howard-Bury, there was a positive rash of sightings by Europeans, most notable being the case recorded by one A. N. Tombazi, a member of the Royal Geographical Society of London, while on a photographic expedition to Sikkim. This gentleman observed one through field-glasses for some time; it was grubbing for roots with a stick on the other side of a valley, and later he found its footprints (which were just like those of Shipton's ABSM).

Numerous Russians have also seen ABSMs, quite apart from the one reported in the Pamirs by A. J. Pronin of Leningrad Uni-iversity, which caused so much excitement in 1957. I can quote but one example: that of Prof. V. K. Leontiev, Chief of the Conservation Department of the Dagestan A.S.S.R., which lies between the Caucasus Mountains and the Caspian Sea.

While on a routine reconnaissance of one of the enormous game reserves in his territory, this experienced field naturalist saw one of the local ABSMs called there a Kaptar, and observed it at a range of from only 50 paces until it disappeared ahead of him seven minutes later about half a mile ahead. His description is completely scientific and most detailed, and he took accurate scale drawings of the imprints it left. It was about seven feet tall, clothed in shaggy hair, had very wide, stubby feet with widely spread toes and an enormous big toe. Its head was small above the ears. It was stoop-shouldered and had a rolling, shambling gait, but when Professor Leontiev fired a shot at its feet, it waltzed about and then made off up a very steep slope with incredible speed. The full report is some 40 pages long and a masterpiece of Russian dévotion to detail.

Detailed as the Russian accounts are, they are as nothing to those recorded by Mongolian scientists. Unfortunately it would be worthless repeating these because, in our lofty western manner we consider anybody living in the area east of Russia as what we choose to call "natives" and anything they, like our American Indians, Africans, and others say, we discredit. Let me therefore turn to the account of a Hollander of higher education, which was published in a scientific journal in Java.

This gentleman's name was Mienheer van Herwaarden, and the incident occurred in 1923, in an area surrounded by rivers called Poeloe Rimau, in the province of Palembang, on the island of Sumatra. Van Herwaarden had been hunting wild pigs and "gone to bush" to await their appearance at a feeding ground. Something in an isolated tree caught his notice and, going to look, he saw clinging to the trunk a creature covered with thick black fur and with a considerable mane depending from its head and running down its midback. After observing it from only a few feet he started to climb the tree but the creature immediately moved upwards. After talking soothingly to it but getting no response, he tried bolder tactics and again started climbing, but this time the creature scrambled out on to a limb that sagged with its weight and then it dropped about ten feet to the ground and started running away. Van Herwaarden raised his rifle and had it in his sights when it was 30 yards away but then, he says, he could not press the trigger because the thing was obsolutely human but for its fur and mane—and it was a female. Its mate was by this time also calling from the nearby forest.

One further case will lay to rest the fatuous statement that nobody, let alone a "white man" has ever seen an ABSM. This occurred to an American long resident in Canada, named William Roe, in the year 1955, near Tete Jaune Cache in Alberta on a peak named Mica Mountain. Mr. Roe was taking a lone hunting trip, he having spent a lifetime in the wilds and being fond of observing animals and doing a little hunting. When at high altitude in a mixed coniferous and broad-leafed bush forest he came upon what he at first thought was a grizzly bear at about 20 paces, feeding on berries by pulling the branches of a bush and stripping the berries with its other hand or paw. This surprised him but then the thing turned and he saw that it was a huge, humanoid female, clothed in short, thick fur. They stared at each other and he raised his rifle but, like all the others, could not press the trigger. The ABSM shambled off and, throwing its head back, gave out a strange half-yell-half-laugh. Roe followed it up and observed it on a nearby ridge; he then searched about and says that he found a place where it had slept and eaten various vegetable materials.

Combined with the numerous other reports of sightings of this type of ABSM, one has no reason to doubt this story. It is quite detailed in the original and makes a number of points that are exactly in accord with what all the others have stated. Among these are two medical doctors in California four years ago returning from an emergency late at night to a place named Redding at the head of the Sacramento Valley. Seeing what they took to be a person sitting by the roadside, they slowed down and dimmed their lights with a view to offering a lift.

Suddenly the "thing" leaped up, took the road in two strides and crashed into the thick bush! Almost exactly the same thing was reported a year later by two hunters on the road where the first footprints occurred in 1958, and I have literally dozens of others from all sorts of people, including a young lady, now 21, who says she met one in the morning mists a little distance from where she was camping with her parents when she was 10 years old.

So, you may well say, people all over the world say they have seen or encountered these creatures, why have not they, or we, captured one? This is also a very fair question, so I will give you another reasonable answer . . . we have.

Such a statement, of course, calls for full documentation. Here it is, starting with the first record we have of such a capture on our own continent—and in southern British Columbia, Canada, no less; and not 100 miles from the United States border.

The particular incident occurred on the morning of July 3, 1884, on the railroad track bordering the Fraser River, near a small place called Yale, which is not 100 miles from the great city of Vancouver and only 20 from the long-inhabited shore of Harrison Lake. It may be called "The Jacko Affair." I herewith quote it in full from a Victoria, B.C. newspaper named the Daily British Colonist.

"Yale, B.C., July 3, 1884—In the immediate vicinity of No. 4 tunnel, situated some 20 miles above this village, are bluffs of rock which have hitherto been unsurmountable, but on Monday morning last were successfully scaled by Mr. Onderdonk's employes on the regular train from Lytton. Assisted by Mr. Costerton, The British Columbia Express Company's messenger, a number of gentlemen from Lytton and points east of that place, after considerable trouble and perilous climbing captured a creature who may truly be called half man and half beast. 'Jacko,' as the creature has been called by his captors, is something of the gorilla type standing about 4 feet 7 inches in height and weighing 127 pounds. He has long black strong hair and resembles a human being with one exception, his entire body, excepting his hands (or paws) and feet are covered with glossy hair about one inch long. His forearm is much longer than a man's forearm, and he possesses extraordinary strength, as he will take hold of a stick and break it by wrenching or twisting it, which no man living could break in the same way. Since his capture he is very reticent, only occasionally uttering a noise which is half bark and half growl. He is, however, becoming daily more attached to his keeper, Mr. George Telbury, of this place, who proposes shortly starting for London, England, to exhibit him. His favorite food so far is berries, and he drinks fresh milk with evident relish. By advice of Dr. Hannington, raw meats have been withheld from Jacko, as the doctor thinks it would have a tendency to make him savage.

The mode of capture was as follows: Ned Austin, the engineer, on coming in sight of the bluff at the eastern end of the No. 4 tunnel saw what he supposed to be a man lying asleep at close proximity to the track, and, as quick as thought, blew the signal to apply the brakes. The brakes were instantly applied, and in a few seconds the train was brought to a standstill. At this moment the supposed man sprang up, and uttering a sharp quick bark be-

155

gan to climb the steep bluff. Conductor R. J. Craig and express messenger Costerton, followed by the baggage man and brakesmen, jumped from the train and knowing they were some 20 minutes ahead of time, immediately gave chase.

After 5 minutes of perilous climbing the then supposed demented Indian was corralled on a projecting shelf of rock where he could neither ascend nor descend. The query now was how to capture him alive, which was quickly decided by Mr. Craig, who crawled on his hands and knees until he was about 40 feet above the creature. Taking a small piece of loose rock he let it fall and it had the desired effect of rendering poor Jacko incapable of resistance for a time at least. The bell rope was then brought up and Jacko was now lowered to terra firma. After firmly binding him and placing him in the baggage car, 'off brakes' was sounded and the train started for Yale. At the station a large crowd who had heard of the capture by telephone from Spuzzum Flat were assembled, and each one anxious to have the first look at the monstrosity, but they were disappointed, as Jacko had been taken off at the machine shop and placed in charge of his present keeper.

The question naturally arises, how came the creature where it was first seen by Mr. Austin? From bruises about its head and body, and apparent soreness since its capture, it is supposed that Jacko returned too near the edge of the bluff, slipped, fell and lay where found until the sound of the rushing train aroused him. Mr. Thomas White, and Mr. Gouin, C.B.E., as well as Mr. Major, who kept a small store about half a mile west of the tunnel during the past 2 years, have mentioned having seen a curious creature at different points between Camps 13 and 17, but no attention was paid to their remarks as people came to the conclusion that they had either seen a bear or stray Indian Dog. Who can unravel the mystery that now surrounds Jacko? Does he belong to a species hitherto unknown in this part of the continent or is he really what the train men first thought he was, a crazy Indian?"

Now, whatever you may think of the press, you cannot just simply dismiss everything reported by it that you don't believe in. Further, this report is excellent, being factual, giving names that were obviously carefully checked even to titles such as the C.B.E. of Mr. Gouin, and hardly being at all speculative. In fact, it is really a model report and one that some modern newsmen might well emulate. Then, the persons concerned were not a bunch of citizens with names only to identify them; they were mostly people with responsible positions who must have been widely known at that time throughout the area, for the railroad played a very important part in the opening up and development of lower British Columbia. The reporter, moreover, himself took a very commonsense view of the business when he inquired what manner of creature this might be and stated flatly that it was competely human but for being covered with silky black hair and having exceptional strength in its arms.

Unfortunately, following the excellent report the news on "Jacko" is pretty slim. The creature was held in captivity for some time, but there is no record of his ever having been examined by scientists. He was simply accepted as an odd event in a world in which odd events were happening all the time. Perhaps some part of him has been preserved and is lying in somebody's attic, or even in a museum. It's happened before.

There have been numerous other reports of captures, from all over the world; I have over 50 on file. None, however, is as plain as the case of poor little "Jacko"—outside of Russia, Mongolia and China, that is. To give these even in brief would call for a large volume, so I quote but one that has for long seemed to me to be outstandingly straightforward. This case comes from official records of the Soviet Army Medical Corps to the Special Commission appointed by the Russian Academy of Sciences to investigate ABSMery, under Professors Porshnev and Shmakov. The incident occurred in 1941, and was put on record by one Lt. Col. V. S. Karapetyan. It states, in his own words:

"From October to December of 1941 our infantry battalion was stationed some 30 kilometers from the town of Buinaksk (in the Dageston, U.S.S.R.). One day the representatives of the local authorities asked me to examine a man caught in the mountains and brought to the district center. My medical advice was needed to establish whether or not this curious creature was a disguised spy.

"I entered a shed with two members of the local authorities. When I asked why I had to examine the man in a cold shed and not a warm room, I was told that the prisoner could not be kept in

157

a warm room. He had sweated profusely in the house and they had to keep him in the shed.

"I can still see the creature as it stood before me, a male, naked and barefooted. And it was doubtlessly a man because its entire shape was human. The chest, back, and shoulders, however, were covered with shaggy hair of a dark brown colour. (It is noteworthy that all the local inhabitants had black hair.) This fur of his was much like that of a bear, and 2 to 3 centimeters long. The fur was thinner and softer below the chest. His wrists were crude and sparsely covered with hair. The palms of his hands and soles of his feet were free of hair. But the hair on his head reached to his shoulders, partly covering his forehead. The hair on his head, moreover, felt very rough to the hand. He had no beard or moustache, though his face was completely covered with a light growth of hair. The hair around his mouth was also short and sparse.

"The man stood absolutely straight with his arms hanging, and his height was above the average—about 180 cm. He stood before me like a giant, his mighty chest thrust forward. His fingers were thick, strong, and exceptionally large. On the whole, he was considerably bigger than any of the local inhabitants.

"His eyes told me nothing. They were dull and empty—the eyes of an animal. And he seemed to me like an animal and nothing more.

"As I learned, he had accepted no food or drink since he was caught. He had asked for nothing and said nothing. When kept in a warm room he sweated profusely. While I was there, some water and then some food (bread) was brought up to his mouth; and someone offered him a hand, but there was no reaction. I gave the verbal conclusion that this was no disguised person, but a wild man of some kind. Then I returned to my unit and never heard of him again."

Yet I know that you will still be saying "But, why haven't we got one?" There are several reasons. First, the vastness and impenetrability of the areas where these comparatively rare creatures live. Secondly, the fact that for the most part being humanoids, if not full men, they possess both a degree of what we call intelligence and a goodly quota of what we call animal instincts. Even primitive peoples are often uncanny in their ability to keep out of sight and their senses are unbelievably acute. All of this renders even a chance encounter quite unlikely.

But the main reason is that up until fairly recently we have never gone about the problem of finding one with much knowledge or common sense. We have looked for them in the wrong places, and we have gone about it in the wrong way. We are now, I hope, going about it in the right way, and I have every reason to believe that we will be successful. I, for one, am looking forward with a good deal of pleasure to seeing what the "experts" have to say when they come face-to-face with one of the thousands of "Abominable Snowmen" which are living today on our mysterious planet.

—Ivan T. Sanderson

Throughout this book. I have quoted Ivan Sanderson a number of times. You might say Mr. Sanderson has been the inspiration for this book. If this book interests you, we recommend Sanderson's "Abominable Snowman Legend Come to Life," over 500 pages of scientific research on this subject.

This is the story of sub-humans on five continents from the early Ice Age until today. It can be purchased by writing to:

Chilton Brooks
1525 Locust Street
Philadelphia, Pa. 19106

DO 'EXTINCT' ANIMALS STILL SURVIVE?

A Closer Look Into the World's Unexplored Jungles and Wastelands May
Reveal Some Strange Creatures, Says a Noted Biologist

By EVERETT H. ORTNER

Reprinted courtesy of Popular Science Monthly—1959

By Popular Science Publishing Co., Inc.

Known to science only by this 1920 photo, Ameranthropoides loysi was seen just once

Does the hairy mammoth, ancient cousin of the elephant, still nip mosses in some frozen Siberian fastness that man has not yet explored? Does the giant moa, a flightless bird that stood 12 feet high, still lay its eggs, big enough to make omelets for 70 people, deep in the New Zealand forests? Is the Impossible Waitoreke possible?

These things may well be, says one eminent zoologist, who considers that reports of the demise of many beasts that roamed the world eons ago are greatly exaggerated, and that still others, whose rumored existence is scoffed at by scientists, are nonetheless alive and waiting to be discovered and stuffed. The expert: Dr. Bernard Heuvelmans, a Belgian whose book setting forth these and other claims has just been translated into English under the title, On the Track of Unknown Animals (Hill and Wang, Inc., New York; $6.95).

The orang pendek is said to speak an unintelligible language.

Consider the Komodo Dragon, says Dr. Heuvelmans. Back in 1912, an airman who made a forced landing on the Malay island of Komodo reported that he had seen fierce dragons there that ate—according to the natives—pigs, goats and deer, and even attacked horses. Zoologists ignored the report when it appeared that the unlucky flier had no university degrees in zoology.

But soon afterward, the Dutch civil administrator obtained the skin of an enormous lizard that matched the "dragon's" description. And shortly after that, several more—evil-eyed, 12-foot monsters—were captured alive, turning the legend of the Komodo Dragon into an undeniable scientific fact.

Or take the case of the coelacanth, a forbidding fossil fish whose obituary, written by paleontologists, had put its time of death at some 70,000,000 years ago. In 1938, one swam, seemingly out of the Mesozoic era, into a South African fisherman's dragnet. Since 1952, flicking fresh salt spray on the scientists' wounds, a number have been captured alive.

What are the chances of an animal's being unknown to science?

161

Less than in the past, but still excellent, according to Dr. Heuvelmans. His reasoning: Two centuries ago, when animals first began to be classified systematically, only 808 amphibians, reptiles, mammals and birds were known. By the beginning of the present century, the number had increased to more than 21,000. By 1960, Dr. Heuvelmans estimates, 60,000 creatures will have been described and named. Since the beginning of the century, he says, approximately 15 previously unknown reptiles or amphibians, 220 mammals and 400 birds have been described each year.

Mummified remains of the moa have been found. But no one knows whether this 12-foot bird still exists.

"This does not mean that they were all really new," he adds. "Many zoologists create new species and subspecies on the strength of barely perceptible differences. All the same, we can reckon that one in 10 of these descriptions refers to a clearly distinct species. Therefore each year the catalogue includes some 40 birds and 20 mammals that were hitherto unknown." Most of these are small songbirds, little rodents or bats, or even small marsupials or insect eaters. "Reptiles and amphibians are much rarer and almost an exception," he says. "There are no more than one or two a year."

Where are these unknown creatures to be found? All over, says Dr. Heuvelmans, but especially in those areas of the world that are little known or even unexplored. Excluding Antarctica, these make up a tenth of the world's surface. The central and northern parts of Greenland, for example, have never been trod by a white man. Much of Africa is known only by aerial photographs. In Asia, there are the Himalayas and much of Siberia, China, Burma, India. The Arabian desert of Dahana is, according to one geographer, "the least explored in the world—infinitely less well-known than the poles."

"Nothing whatever is known about the mountainous center of New Guinea," says Dr. Heuvelmans, "except that various tribes, many of them pygmies, still live there in the Stone Age. In 1938 the American naturalist Richard Archbold accidentally discovered in the western part of the island an excellently irrigated valley inhabited by 60,000 people. And in June, 1954, patrol aircraft found quite unknown tribes in valleys in the southwest. Their population was estimated at 100,000, a third of the number of Papuans already known."

Australia? Most of the interior is covered with deserts of sand, salt and thorny bush. "Hardly anybody goes there," says Dr. Heuvelmans, "except a few prospectors, who come back with tales of animals so fantastic that they are usually thought to be drunken visions." There are even whole mountain ranges that have never been seen except from airplanes.

But the continent that holds the most mysteries is surely South America, with its steaming, tropical jungles—2,000,000 square miles for the Amazon basin alone. Where the Orinoco River rises, in one corner of Venezuela, are vast limestone plateaus—mesas—3,000 to 10,000 feet above the surrounding jungle. Some of these are almost 20 miles long, great islands of thick vegetation about which almost nothing is known. On one of these is a waterfall 15 times as high as Niagara. It was discovered only 20 years ago.

Strange legends hang over all these unexplored areas. One trav-

A survival from the Mesozoic era of 70,000,000 years ago, the coelacanth had long been written off by scientists as extinct. But in 1938 one was caught in South Africa, and since then many more of the fossil fish have swum into scientists' hands.

163

A "living fossil" that carries its age well, the giant tortoise of the Galapagos Islands has remained almost unchanged for 190,000,000 years.

eler in Africa, Frank H. Melland, heard from the natives of Northern Rhodesia of a fierce creature that lived in the nearby Jiundu swamp—like a bird, but not exactly a bird; more like a lizard with wings of skin like a bat's.

Melland noted this down, but only later did he realize its hair-raising implications. Then he renewed his questioning. The beast's wing span, they said, was between four and seven feet; it had no feathers at all; its skin was bare and smooth; its beak was full of teeth.

Melland was staggered. What he had was a complete description of a pterodactyl — a giant flesh-eating flying dragon known only to paleontologists, and supposedly extinct for tens of millions of years.

When Melland showed the natives pictures of a reconstruction of a pterodactyl, they nodded and muttered excitedly: "Konamato!"

Other travelers also heard of the strange flying beast, but only one of them—Ivan T. Sanderson, leader of an expedition into West Africa—ever saw it. Working along a river, he shot a fruit-eating bat, went into the water to recover it. Suddenly his companion shouted: "Look out!"

"And I looked," Sanderson later reported. "Then I let out a shout also and instantly bobbed down under the water, because, coming straight at me only a few feet above the water was a black thing the size of an eagle. I had only a glimpse of its face, yet that was quite sufficient, for its lower jaw hung open and bore a semi-circle of pointed white teeth set about their own width apart from each other."

Across the Sunda Straits from Java, where 65 years ago were dug up the skull fragments that scientists have assigned to a creature they call pithecanthopus erectus—apeman who stands up—natives speak of another apeman, believed to exist today. The Dutch settlers call him "orang pendek" (little man) or "orang letjo" (gibbering man).

A small creature, between 2½ and five feet tall, the orang pendek is said to speak an unintelligible language. Reports say that its skin is pinkish brown, and covered all over with short dark hair. It has no tail. It walks on the ground. It is known, says Heuvelmans, all over southern Sumatra below the Equator, and has been described many times.

Perhaps the man to get nearest to one was a Dutchman named van Herwaarden, a curious man who climbed a tree to study, at close range, one of the creatures hiding there.

"The very dark hair on its head fell just below the shoulder blades or even almost to the waist," he reported later. "The brown face was almost hairless, whilst its forehead seemed to be high rather than low. The eyes were frankly moving; they were of the darkest color, very lively, and like human eyes. This specimen was of the female sex and about five feet high.

"There was nothing repulsive or ugly about its face, nor was it

Could this ugly snout be mistaken in the dark jungle? The mandrill, a fierce African baboon, may be the reality behind native tales of a strange and dangerous bear.

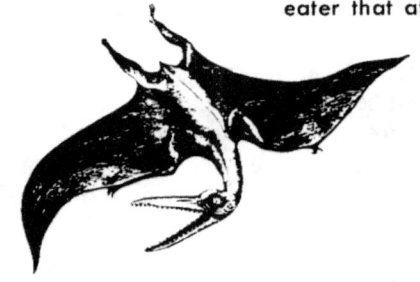

The pterodactyl: Was it this flying flesh-eater that attacked one explorer?

at all apelike, although the quick nervous movements of its eyes and mouth were very like those of a monkey in distress. I began to talk in a calm and friendly way, but it did not make much difference. When I raised my gun to the little female I heard a plaintive 'hu-hu,' which was at once answered by similar echoes in the forest."

When the creature, eluding van Herwaarden, slid down the tree and fled, the Dutchman raised his rifle to fire. But: " Many people may think me childish if I say when I saw its flying hair in my sights I did not pull the trigger," he said later. "I suddenly felt that I was going to commit murder."

Could an anthopoid—a large, tailless ape with long arms—exist in the Americas? Scientists doubted it, until in 1920 a Swiss geologist, Francois de Loys, brought back evidence from the Sierra de Perijaa, a mountain range that bestrides the Colombian-Venezuelan border. Attacked by two ferocious giant monkeys, five feet tall, de Loys' party killed one, propped it up on a crate, and photographed it (page 84). Scientists are still puzzling over how to classify it.

Many other shadow creatures pace through Dr. Heuvelman's bestiary of unknown animals. The neomylodon, a cousin of both the sloth and armadillo, is one. Here there is more to go on than just native rumors. The animals—big as oxen—have left, curiously preserved in an Argentinian cave, skeletons, droppings, even wisps of the hay they fed on. And there are several pieces of hide.

Apparently the neomylodon was on its evolutionary way to developing an armor plate like that of its cousin the armadillo, for it is a strange hide, of thickish leather with bony nodules—like pebbles—embedded in it. And here scientists disagree. Are these bits of hide from an animal that lived recently, or were they mummified by the cave's atmosphere? Members of the London Zoological Society, who examined a piece in 1899, thought it fresh: The leather was supple, there were remains of muscles and ligaments, and a "coating of dried serum (blood) was still preserved on the cut edges." But this was 60 years ago, and, to date, no one has brought home a neomylodon.

Nor has anyone brought back alive one of the super-anacondas of Brazil—said to run to 150 feet in length. But Father Protesius Frickel, a Franciscan priest at Oriscima, Brazil, studied one and reported that its eyes were "as large as plates." And another, an estimated 115-footer, was photographed as recently as 1948, after machine gunners had spent 500 bullets killing it. Its picture taken, the giant snake was casually rolled back into the river whence it had come. No one knew of its rarity.

The natural deep-freeze of Siberia has enabled scientists to study the mammoth more closely, perhaps, than any other fossil creature. Several have been found, perfectly preserved as they were in life many thousands of years ago, in giant ice cubes.

No one really knows what caused the mammoth to disappear. As the Ice Age melted into time and the glaciers retreated, the mammoths — adapted by their long, hairy coats only to cold — were backed up into the Siberian tundras. And then they were seen no more.

But Heuvelmans thinks they may just still be around—but in the forests instead of on the plains. He puts a finger on the map where the climate could still support them: It is the unexplored icy heart of the taiga—the Northern Siberian forests, largest in the world.

An analysis of the stomach contents of mammoths, he points out, shows that in winter they browsed on the leaves and branches of conifers, arctic willow and other dwarf northern trees—exactly the flora that one finds in the taiga.

Do we know for sure, Heuvelmans asks, that they are not still alive in the taiga? Are we sure that those frozen carcasses were in cold storage as long as some scientists say?

Heuvelmans has similar doubts about the demise of the moa, the huge bird that stood—or stands—twice as tall as a man. While no one has reported seeing any moas around for quite a while, several facts support the possibility that they still linger on. There are the native traditions, which give a very detailed picture of the bird. Could they describe a moa unless their not-too-remote ancestors had seen one? Also, on South Island, New Zealand, have been found comparatively fresh remains: mummified remnants of the body and skin, and brownish feathers. Shells have been found, too, of eggs over 10 inches long.

There is the case of the Bermuda petrel, which science had considered extinct—until, in 1951, Dr. Robert Cushman Murphy of the American Museum of Natural History caught, banded and released five live ones. It may well be like that with the moa, says the hopeful Dr. Heuvelmans.

New Zealand may also conceal a small animal that would, if it exists, effect a revolution in scientific theory. It is the waitoreke—called "impossible" by paleontologists, "existing" by the native Maoris.

If anyone finds one of the creatures, especially a female, Dr. Heuvelmans would like him to examine it closely and let him know whether it has teats. If it has, many textbooks will have to be junked. For teats would make the waitoreke a mammal—the only one in New Zealand—and all our geographical theories about how and when New Zealand and Australia were connected will have to be revised. "The date of submergence of certain continental bridges would have to be changed by several tens of millions of years," Dr. Heuvelmans says—if the Impossible Waitoreke has teats.

How much skepticism should one bring to native legends? Capt. William Hichens, an Englishman who had expressed disbelief in the existence of a great African snake, the "lau," writes:

"Mshengu lived on the southeast side of the Wembare swamps and, as a young man, had traveled and hunted around and over Victoria Nyanza and the Nile swamps. To my suggestion that there was no lau, he said, 'I might have said, as a young man, that there is no such thing as a motor car. I had never seen or heard of one then. But there is your motor car in the sight of my eyes and I have sat on its chairs and heard its bowels digest inside it. It is thus of the lau.' "

Still doubtful? Consider this: In 1941, in Sanquin, China, a scientist by the name of Weidenreik uncovered giant teeth and parts of a huge skull. Research indicated that the remains were at least 500,000 years old. After they were studied carefully, it was calculated that they belonged to a huge humanoid-type creature that would have had twice the bulk of a Gorilla.

The Yakima Daily Republic, Nov. 2, 1960

EARTH, STARS AND MAN . . . APE MEN AND GIANTS

By DON OAKLEY and JOHN LANE

APE MAN OF JAVA —ONE OF THE EARLIEST KNOWN MEN

SKULL OF JAVA APE MAN WAS RESTORED BY FRANZ WEIDENREICH. PEKING MAN IS QUITE SIMILAR AND PROBABLY CLOSELY RELATED.

Near the end of the 19th century, a young Dutch doctor, Eugene Dubois, had a hunch. Java might be the place to hunt for the original man, for that land had once been connected to the Asiatic mainland and had escaped the ice ages.

Amazingly enough, in 1891, Dubois actually found the fossil he had dreamed of—the skullcap of a manlike creature never before seen. A year later he found a complete human thighbone near the original find. On the basis of this, he named it Pithecanthropus erectus—the erect ape man.

It was the same old story: scientists were skeptical. What did this amateur know? The skull and thigh obviously did not belong together. As a result, Dubois put his ape man away in a strongbox. It was not until 1923, when opinions had changed, that he allowed him to be viewed again.

The Java man was eventually redeemed by discovery of a related type near Peking, China. From 1929 until World War II, an international team carried out excavations. Tragically, except for plaster casts, all the remains of this vastly important Sinanthropus pekinensis—China Man of Peking—were lost during the evacuation after Pearl Harbor.

One of the strangest mysteries in the whole story of man concerns giants. In 1935, a young paleontologist, G. H. R. von Koenigswald, purchased three gigantic humanlike teeth in a Chinese drug store in Hong Kong. (The Chinese had used powdered fossil bones as medicine for generations.) Hesitantly, he coined the name Gigantopithecus—giant ape.

Later in Java, von Koenigswald found a skull identical to Dubois' original ape man, further strengthening that creature's place in history (now dated as living 500,000 years ago). But he also unearthed several jaw fragments of two different, huge beings which he called "Robust Ape Man" and "Big Man of Old Java." The latter apparently had twice the bulk of a gorilla. There was no longer any doubt that giants—of some kind—had once walked the earth.

The matter is still not settled as to whether these giants were the ancestors of the Java and China ape men, or whether they were strange, freakish side branches. Only continued digging by new

170

PEKING MAN USED FIRE

GIANT'S JAW OF JAVA AND CHINA

MODERN MAN'S JAW

WERE THERE GIANTS IN THOSE DAYS?

generations of scientific detectives can throw new light on this intriguing riddle or, perhaps, unearth something even stranger.

Giants really did exist in the past. We've already seen how creatures that were thought to be extinct for millions of years are still with us today. Why then is the existence of these Giants doubted? They are bigger than most animals, more intelligent and are self sufficient in the mountains and rain forests.

To continue the search for the truth in the matter concerning the existence of Bigfoot, Giant Hairy Ape, and Sasquatch I pushed my efforts into what I consider pre-expeditions to gather more material and to find out the best area to set up the extended Bigfoot expedition.

During 1966 I discovered new evidence of the existence of the Giant Hairy Ape around the Mt. St. Helens area in Washington state. My son, Clint, and I have made many trips into this area, the last of which was a horseback excursion into Ape Canyon. Here is what happened:

We set up camp in the lower part of Ape Canyon and in the days that followed, scouted half way up both sides of the canyon without success. As we loaded the horses in the truck and headed down for Woodland, to leave, I decided to stop at Charley Erion's place and see if he had turned up any new traces of the hairy giants.

We found Charley at home and were delighted to learn that he and his son, Jim, had found fresh tracks four nights in a row in a plowed field on the lower part of the ranch. We quickly saddled up and rode out there. We came upon tracks everywhere . . . tracks only a few days old!

We could see where the creature had come down from the foothills, stepped over a four-foot fence, walked across the plowed field, stepped over a five-foot gate into an alfalfa field. Eventually, it had come back over the gate, ate some early berries, then returned to the hills. Charlie said that his dogs had barked all four nights and that the cattle had all come up close to the ranch outbuildings.

I made plaster casts of the tracks and they measured 18 inches long and 6¾ inches across the ball of the foot. The tracks were absolutely the real thing! No one could have faked them!

Charley said he had also found tracks up on the side of Ape Canyon in August of 1965 and a friend had told him he was sure he had seen a huge creature that wasn't a bear up in the canyon about that same time. Charley had been logging in that area for some months and he was sure there were giant hairy apes in the surrounding forest at that time.

Prentis Beck and I went back on another pre-expedition to see if

172

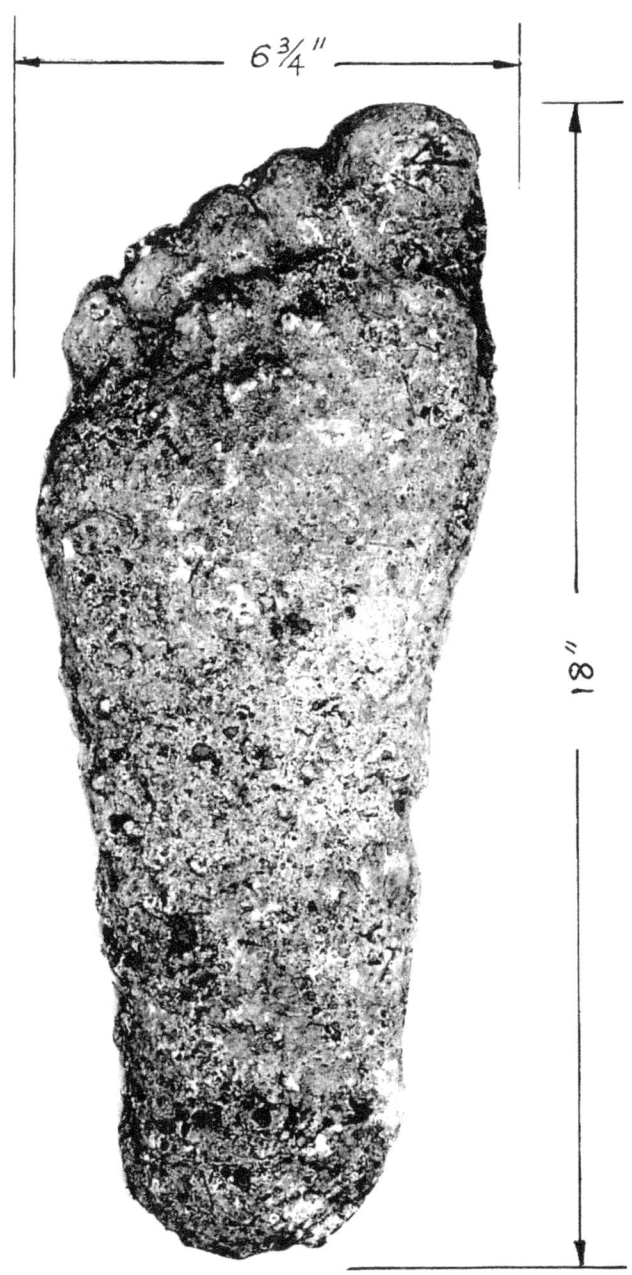

6¾"

18"

Cast of giant hairy ape track made July 7, 1966 by the author at the
Charlie Erion Ranch near Woodland, Washington.

we could locate new tracks and possibly even sight one of the Giant Hairy Apes. We decided to camp at the upper end of Smith Creek which is about four miles from Spirit Lake. We camped on the canyon wall of Smith Creek and looked directly east toward Mt. Adams and the bright evening sun as it shone through the trees and the surrounding forest gave the trees the appearance of fluorescent green. As I looked out over the canyon, smoke curled up from far down below in the canyon and drifted lazily over the evening air. While we sat and took in this breath-taking scene, we couldn't help but feel how free and beautiful these giants' world is in this country, but we soon got back to the reality and purpose of our being here and dug into our night chores of getting our beans and bacon cooked. We went to bed that night with anticipation of tomorrow's adventures. Early the next morning we headed up a winding trail toward Abraham Flats which is located near the head of Ape Canyon. As we progressed mile after mile along the trail, we soon realized that we had come to this spot a little early in the season because the winter's snow covered and prevented further travel on the trail. We came to where we could no longer find the trail, and we made the decision to turn back. It was a disappointment to us as we had looked forward to finding the remains of the old miner's cabin as related previously in the book. When we left camp I told Prentis we should go down into Woodland because I was sure we could still find the track I had made casts of in July, 1966. On our way down off the mountain we stopped at the general store of St. Helens. We had a delightful visit with the owners of the store who told us that they had talked to many people who had reported seeing these creatures. This is always refreshing to us because of the general skepticism on the part of some. We bid our friends goodbye and headed for the Charlie Erions ranch.

We found Charlie home and we all went out to the field where the tracks had been. Charlie explained there had been no more visits since I had been there last. This was a good experience for Prentis as it was the first real tracks that he had seen. Even though they were a few weeks old, the tracks were still deeply imbedded in the soil. Since there was nothing more Charlie could tell us at that time, we decided to head back over the mountains for Yakima, thus ending this pre-expedition into the Giant Hairy Ape area.

About this time I realized the need for someone skilled in the art of tracking. I had thought of many prospects to fill that need but hadn't come up with anyone definite. Then one night I was down to a friend's place, Bob Gimlin, who is a quarter Apache Indian. I knew Bob spent a good deal of time in the mountains and had done much tracking of his own for wild game, and as we talked over the Bigfoot situation, Bob said he would be delighted to try his hand at helping us track these giants. At that time we planned another pre-expedition and decided it would be to try and go through Ape Canyon from the bottom up. We left Yakima on the evening of September 23 with our horses and gear in the truck. As we arrived the next morning, we made a fine camp underneath some gigantic Douglas fir trees. These made excellent shelter as it had been raining some and the warm crackling sound of the fire was a welcome sound and restful after the arduous trip in over treacherous log slashes, dense woods and underbrush to find this good site. We saddled up our horses and decided to go back up on the ledge behind the camp for a look-see before darkness fell. While riding I showed Bob an old-looking trail where trackings had been sighted the previous year. On this particular road there is a side canyon which leads into Ape Canyon and it is extremely steep and brushy. One of the Giant Hairy Apes had managed to climb down its steep sides last year and Charlie Erions and a few other loggers decided not to follow because of the extreme angle.

By the time we made it back to camp it had stopped raining and the night wind filled our lungs with the clean air as only the fresh mountain breezes can. To the rider on the trail, the thought of the welcome campfire and warm meal causes one to hurry toward this respite.

As I cooked our evening meal Bob fed the horses and we settled down to talk over events of the next day and to plan our trip up to Ape Canyon. The black of the night settled around the camp and we began to hear coyotes on the ridge howl their lonesome wails to the sky. Bob had taken his faithful German Shepherd, Ace, along and we were amused at his expressions when he heard the coyotes.

175

The next morning we made for Ape Canyon and we found that the creek was low. The best way to travel is right up the creek, as far as freedom from brush and foliage is concerned, but it is rough for the horses since there is slippery moss on the rocks. As one travels into this almost timeless canyon, the moss-covered trees, dense ferns and other foliage casts a rather eerie atmosphere in the steep creek bottom. About two miles along we came to an abrupt halt because there was a practically impassable log slash through the bottom of the canyon. We decided to try the side of the canyon walls which at that point are not perpendicular, but even this failed, as it is impossible except on foot.

Bob showed me many signs as we traveled the trail through the dense brush—some I could see, others I could not. We decided that rather than leave the horses, we would cut up over the canyon wall and see as much as we could along the edge. We heard a sound very similar to a gruff growl while we climbed the canyon wall, but it was not heard again even though we stopped to listen. The rest of that day we travelled a steep looking road that had just been opened up and led us into fresh virgin timber country. This particular logging trail would have been excellent for finding tracks of evidence of any kind, but even though we travelled about ten miles, no signs were seen.

The following day found the same type of going with the same results. Upon our return to camp, we concluded that our trip should be brought to an end. There were several more trips such as this that were rewarding in information but were not too fruitful in sightings or tracks; however, they bring us closer to that one expedition we eagerly anticipate.

When I returned from my latest pre-expedition, much to my surprise I received a phone call which related an amazing story of a high school boy here in my home town who had come face to face with a gigantic creature west of Yakima. I checked this story thoroughly by interviewing the boy and his father and mother, and it seemed to me an outstandingly straightforward account.

Ken Pettijohn was returning home late at night, Monday, September 19. As he rounded a bend in the road, his light shone on

Sketch by author from description by Ken Pettijohn

what he thought was a huge man covered with silvery white hair standing in the middle of the road. There was a drizzling rain falling, and when Ken saw this creature, he slammed on his brakes and stopped about three feet short of the figure. The creature held up his arm up over his eyes to shield them from the bright lights. In the meantime, Ken's engine stopped because of the suddenness with which he applied his brakes. The creature then walked around the back of the car and then around to the window where Ken was sitting desperately trying to start his engine. The creature stooped down and peered in at Ken. The sensation Ken felt was one of horror, and he was greatly relieved when the engine started and he could get away from there. He could see the creature's silhouette as the lightning lighted up the sky when he looked in his rearview mirror as he drove away. His description of the giant coincided completely with those of previous sightings, even though he did not know of this book or other sightings.

I feel that in Ken Pettijohn's stepping forward and telling his bold story, it may help bring out other stories of incidents by people who have had similar experiences.

What then is the future of the Abominable Snowman of America?

Well, as I said we would at the outset of this book, we've heard the ancient stories of the American Indian, talked with mountain men about these creatures and read eye-witness accounts in magazines and newspapers. The stories we have presented here are true. Bigfoot, Giant Hairy Ape, Sasquatch, Aboriginal Mountain Giant, Abominable Snowman . . . call him what you will . . . he exists. Whether we like it or not he is here with us on this planet, today, in the 20th century. A creature of enormous historical significance. A ghost from a long-dead day that should whet the imagination of the young at heart, the adventurer and the eternal individualist. Their future then, truly belongs to all of us.

Would You Like to Be a Part in This Tremendous Adventure?

Well, here's the chance you have been waiting for! Become a member of "THE ABOMINABLE SNOWMEN CLUB OF AMERICA." Then you'll have a chance to personally take part in what very well may be the GREATEST EXPEDITION IN HISTORY. Here's how . . .

Just print you name and address on the coupon on the back of this page, tear the whole page out and send in for a FULL TWO YEARS MEMBERSHIP. Receive your official 8x10-inch Membership Certificate signed by the Author—Roger Patterson. Also receive 2 GREAT 12-INCH LONG-PLAYING RECORD ALBUMS with over one hour of actual recorded story interviews.

These recordings were made with people who actually came in CONTACT WITH THESE STRANGE GIANTS! You'll hear Jenkins' "Encounter with CALIFORNIA'S BIGFOOT," Fred Beck's "ATTACK IN APE CANYON," William Roe's hunting incident with the "human looking" female SASQUATCH. Also hear Albert Osterman's personal story of how he was KIDNAPPED BY THE SASQUATCH OF B. C. and lived 6 days with them before escaping. These records are available only through the offer in this book.

Any time after receiving your membership certificate, please send a letter stating why you would like to become a member of this Expedition (also stating your qualifications). If you are selected, you will have a special part (paid a salary) in the Expedition.

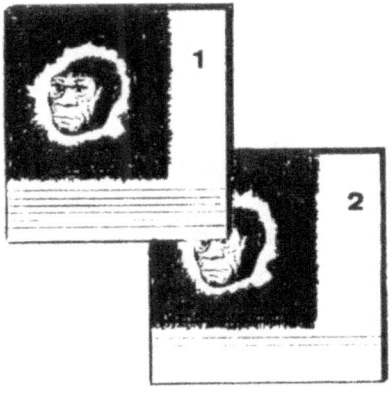

Every four months you will receive a bulletin with the latest news about the ABOMINABLE SNOWMEN OF AMERICA,

179

including all the findings of our planned expeditions into CALIFORNIA AND WASHINGTON'S wilderness areas.

In return, you will send us any information you have on these giant creatures and we will pass it along to the rest of the club.

Sound interesting? YOU BET YOUR BOTTOM DOLLAR IT IS! Send today for your membership, record albums and a chance to be a member of one of the greatest expeditions in history! All this for ONLY $5.95.

Make this the biggest day of your life—write NOW to:

ABOMINABLE SNOWMEN CLUB OF AMERICA
P. O. BOX 836, YAKIMA, WASH. 98901

I would like to become a member in A.S.C.A. Here is my $5.95, plus 50 cents for postage and handling.

NAME .... *DO NOT REPLY. ORGANIZATION DEFUNCT.*

ADDRESS ....

CITY............. ....STATE...........................

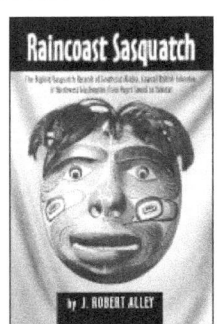